Book 8C

PUBLISHED BY THE PRESS SYNDICATE OF THE UNIVERSITY OF CAMBRIDGE
The Pitt Building, Trumpington Street, Cambridge, United Kingdom

CAMBRIDGE UNIVERSITY PRESS
The Edinburgh Building, Cambridge CB2 2RU, UK
40 West 20th Street, New York, NY 10011-4211, USA
477 Williamstown Road, Port Melbourne, VIC 3207, Australia
Ruiz de Alarcón 13, 28014 Madrid, Spain
Dock House, The Waterfront, Cape Town 8001, South Africa

http://www.cambridge.org/

First published 2003

Printed in the United Kingdom at the University Press, Cambridge

Typeface Minion *System* QuarkXPress®

A catalogue record for this book is available from the British Library

ISBN 0 521 53801 7 paperback

Typesetting and technical illustrations by The School Mathematics Project
Illustrations on pages 158, 168 and 169 by David Parkins
Other illustrations by Robert Calow and Steve Lach at Eikon Illustration
Photograph on page 171 by Paul Scruton
Other photographs by Graham Portlock
Cover image © Image Bank/Antonio Rosario
Cover design by Angela Ashton

The authors and publishers thank Katie Atkinson for her help with the production of this book.

Contents

1 Graphs that tell stories

This is about graphs that show things changing as time passes.
The work will help you

- get information from graphs
- sketch graphs describing real situations

A Into the bath

Peter is taking a bath.
This graph shows the level of the water in his bath.

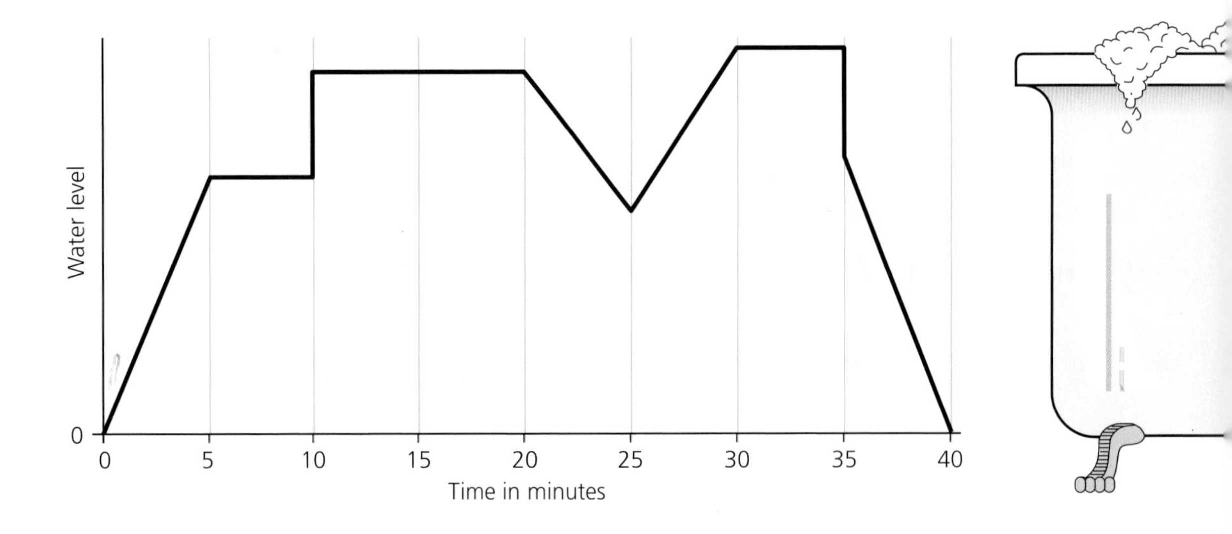

A1 At 10 minutes, Peter gets into the bath.
So the water level rises very quickly.

(a) What is happening between 30 and 35 minutes?

(b) What do you think happens at 35 minutes?

(c) What is happening between 35 and 40 minutes?

A2 Here are three more bath graphs.
Write down what each one tells you.
You can write it as a story if you want.
(Remember to say exactly when the taps go on and off,
and the plug is put in or taken out.)

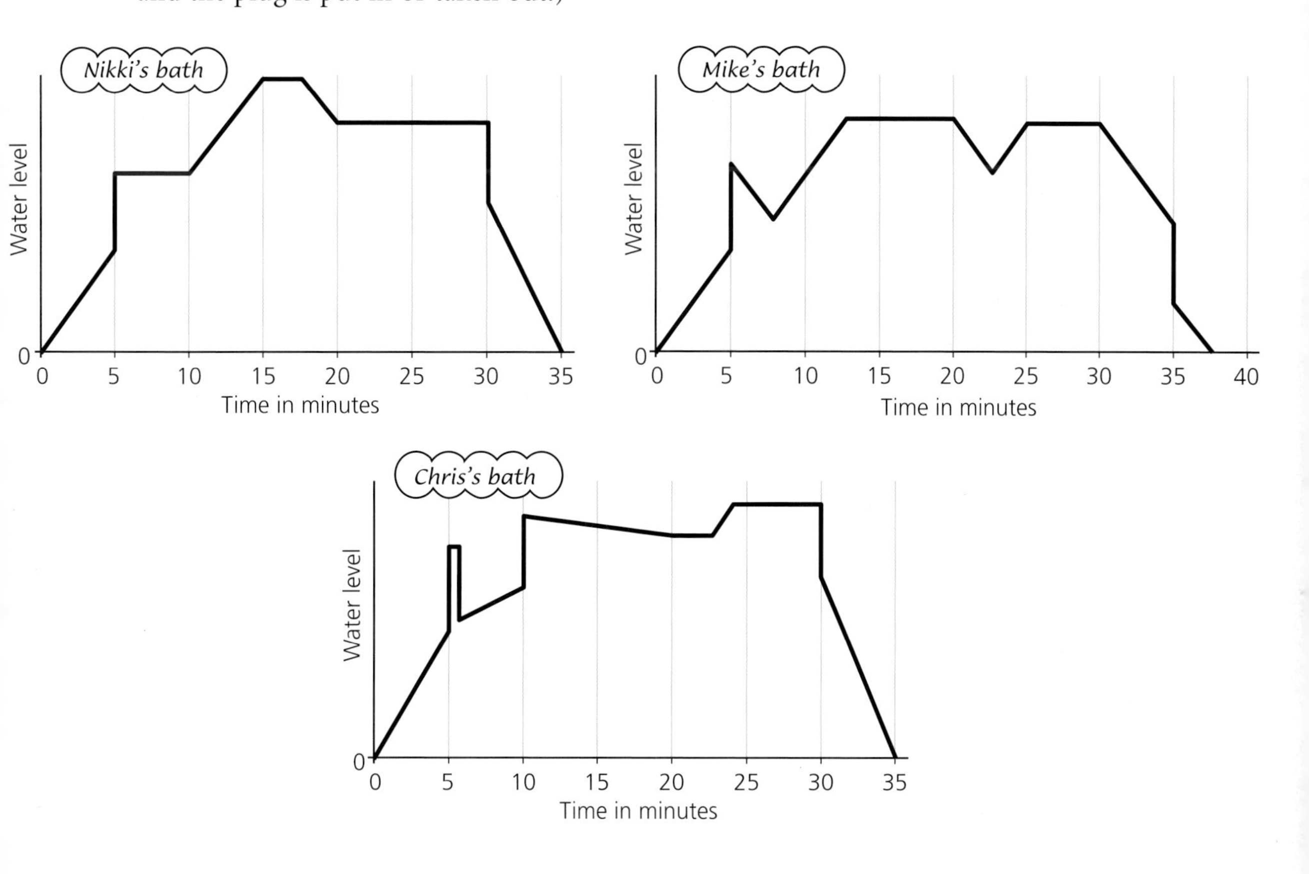

A3 Sketch a graph for this bath story.

I decided to have a long bath.

First I ran the cold tap for 5 minutes,
and then I turned on the hot tap as well.
After 10 minutes there was enough water,
so I turned both taps off and got in.

After I had been soaking for 10 minutes, the phone rang.
I got out and talked to Jane for 5 minutes.

I got back in, but the water was cold, so after another 5 minutes
I got out and let all the water out.

A4 Sketch a bath graph of your own.
Give it to someone else and ask them to write a bath story for it.

B Filling up

Imagine that you are filling a bottle from a tap.
The water from the tap is flowing into the bottle steadily.

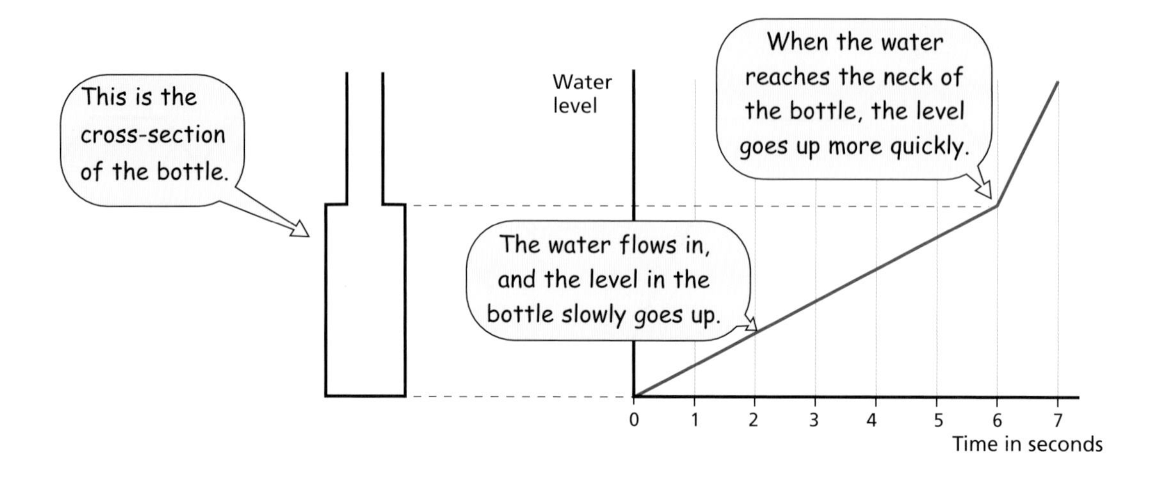

B1 Here is a different shaped bottle.

Which graph shows the level of water in the bottle as it is filled?

B2 Here are three bottles and three graphs.
Imagine each bottle is being filled from a tap flowing steadily.

Which graph goes with each bottle?

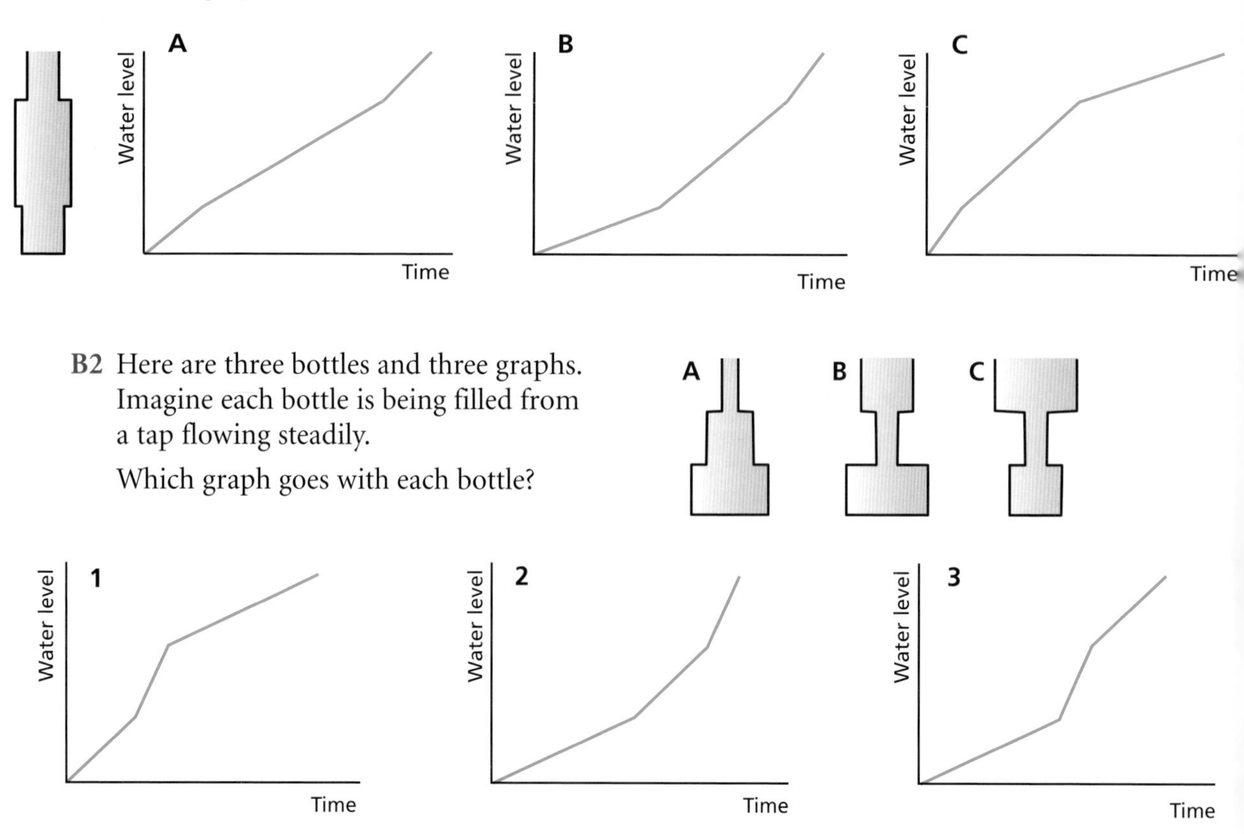

B3 Imagine each of these vases is being filled from a steady tap.
Draw sketch graphs of how the water level changes.

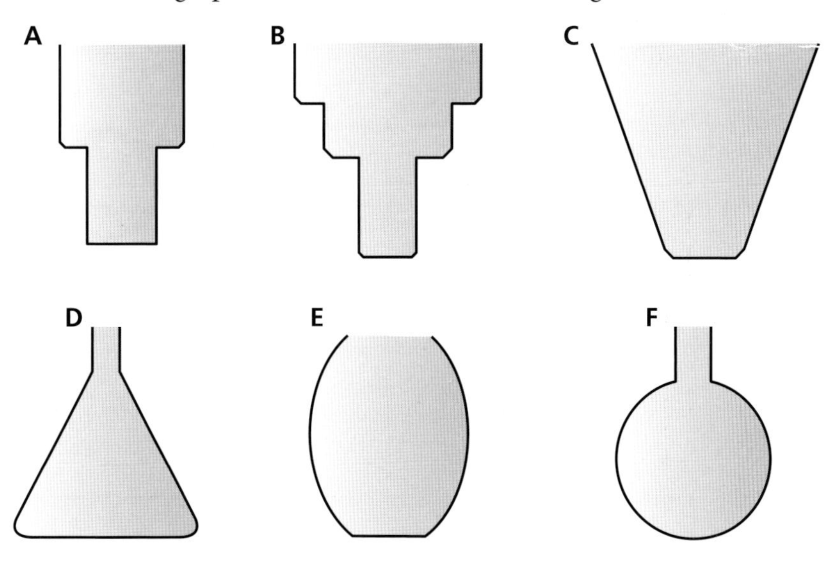

B4 Here are two graphs.
Each shows the water level as a container is steadily filled.

For each graph, sketch a container whose shape fits the graph.

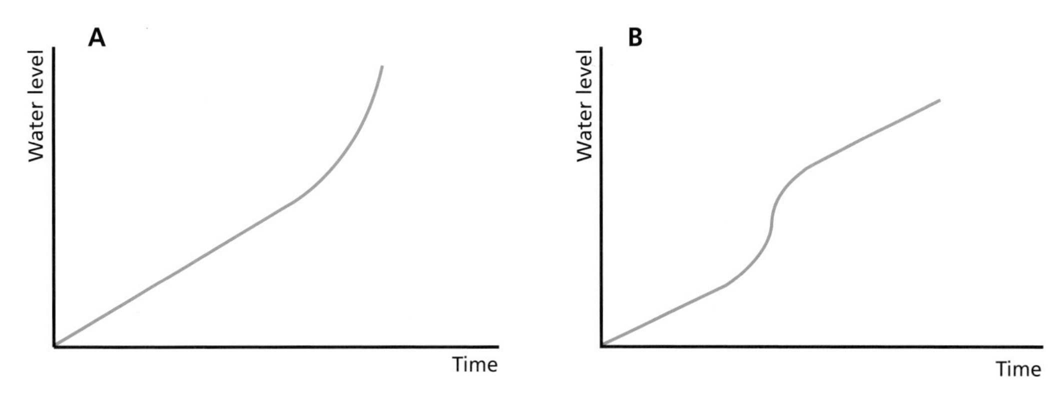

B5 Sketch a water level graph.
Give it to someone else to sketch the container that fits the graph.

Make sure you know the answer yourself!

C Speed

Jude is a keen cyclist.
She always keeps an eye on the speed she is going at.

When she cycles to school, she goes along Clark Road, Beacon Road, Hill View Road and School Road.

This graph shows her speed on the way to school.

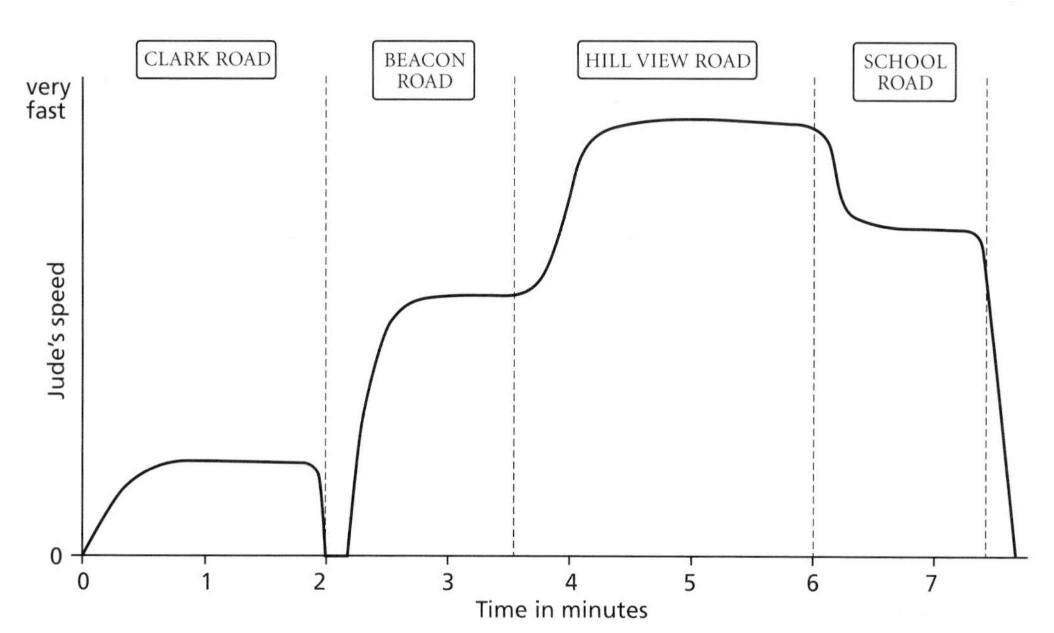

You can see that Jude cycled quite slowly along Clark Road, But she went more quickly along most of Beacon Road.

C1 (a) Along which road did Jude cycle fastest?

(b) Along which road did she cycle second fastest?

(c) There are traffic lights at the end of one road. Jude had to stop at the lights.

(i) Which road do you think the traffic lights are at the end of?

(ii) How many minutes does it take Jude to cycle from home to the traffic lights?

C2 Masud went for a cycle ride one day.
This is a graph of his speed on the ride.

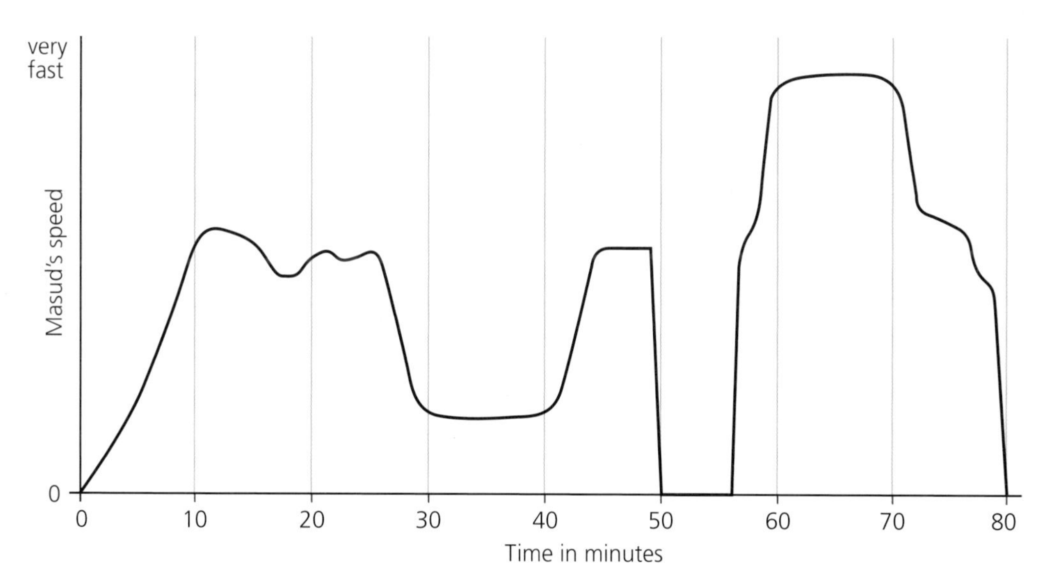

(a) At one time on his ride Masud stopped to talk to a friend.
How many minutes had he been cycling for when he stopped?

(b) For 10 minutes Masud was cycling slowly up a hill.
Between what times was this?

(c) For 10 minutes, Masud was cycling very fast downhill.
Between what times was this?

C3 Callie is a sprinter.
Her trainer has written down instructions for a training run.

Training: Callie

First jog for about 10 minutes.
Then run as fast as you can for 5 minutes,
followed by jogging for 5 minutes.
Then walk for 5 minutes.

Rest sitting for 5 minutes, then jog for 5 minutes
and then sprint for 5 minutes.
Walk for 5 minutes and then you can stop.

See you Wednesday,
Fred

Sketch a graph of Callie's speed as she trains.

Label your up (speed) axis '0' at the bottom and 'very fast' at the top.

Your across (time) axis will need to go from 0 to 50 minutes.

C4 Derek goes for a bike ride.
First he cycles at a steady rate.
Then he goes steadily but slowly up a long hill.
Lastly he goes quickly down a steep hill, and then stops.

(a) Which of the sketch graphs do you think shows his speed?

(b) Make up your own stories for the other two graphs.

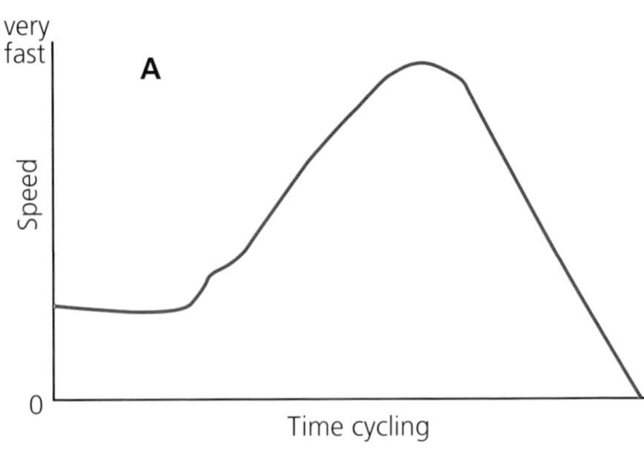

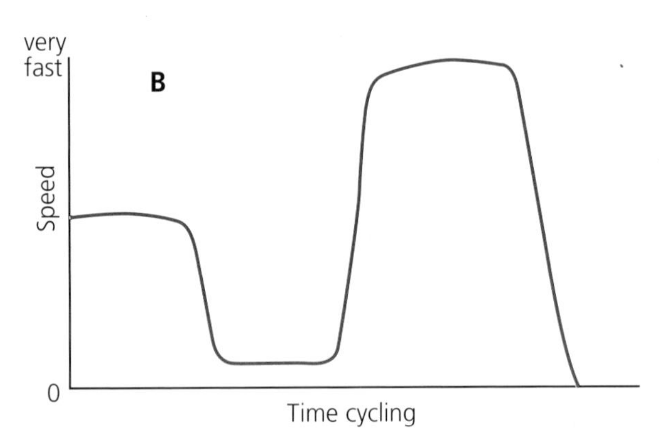

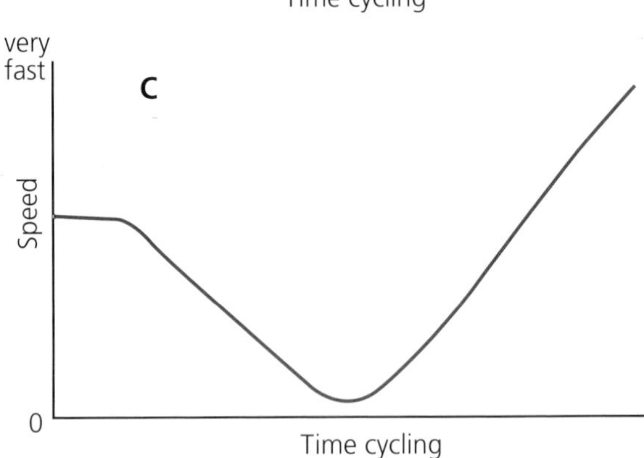

C5 Della is trying out a friend's moped to see if she wants to buy it.

She takes it for a test run and tells her friend about it.

Draw a sketch graph of her speed on the moped.

Well, it went OK for about 5 minutes, and then I got held up in the High Street for 5 minutes.

It took me ages to get up Quarry Bank – about 10 minutes, the hill was so steep. Coming down the other side it went like the wind for about 5 minutes, but then it just conked out completely.

I let it cool off for 10 minutes, but then I couldn't start it and it took me 10 minutes to push it back here!

C6 Make up a speed story of your own.

Draw a sketch graph that fits your story.

What progress have you made?

Statement	Evidence
I can sketch real-life graphs.	**1** Sketch a speed graph for this story.

Went out cycling – I was out for an hour altogether.
First I cycled up Crudge Bank – took me 10 minutes, going really slowly.
Then I stopped for 10 minutes!

But then top speed down the other side for 5 whole minutes!
Then I met Jim walking and pushed my bike alongside him for 10 minutes.

After that, I cycled steadily for 15 minutes, and then sprinted for 10 minutes all the way home.

Statement	Evidence
I can interpret real-life graphs.	**2** Write a story for this graph. It shows the water level when Suhela had a bath.

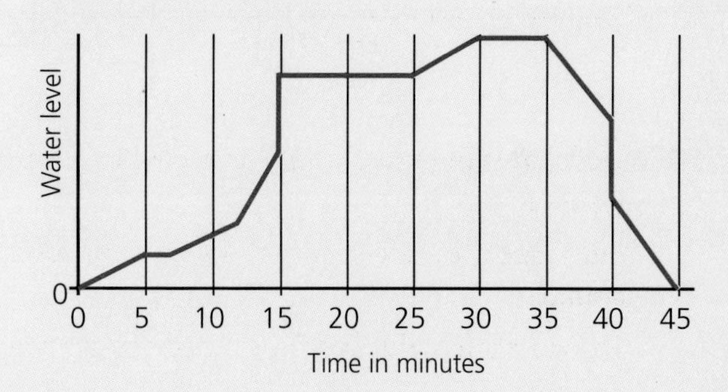

3 Write a story for this graph of a car's speed.

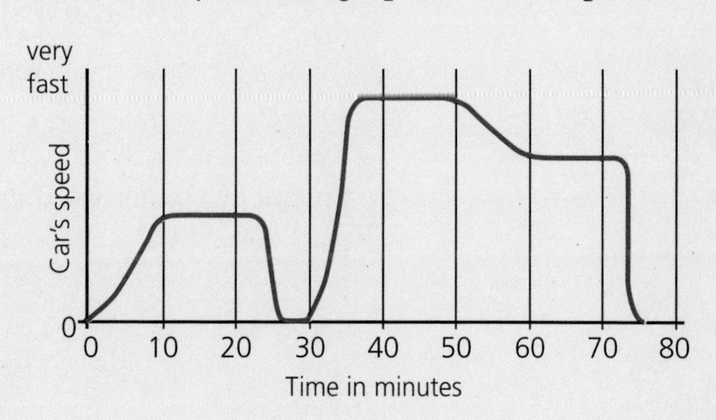

What progress have you made?

Statement

I can interpret and sketch more complex real-life graphs.

Evidence

4 This vase is being filled from a tap that is flowing steadily.

Draw a sketch graph for the water level in the vase as it is filled.

5 Sketch a vase whose shape fits this water level graph.

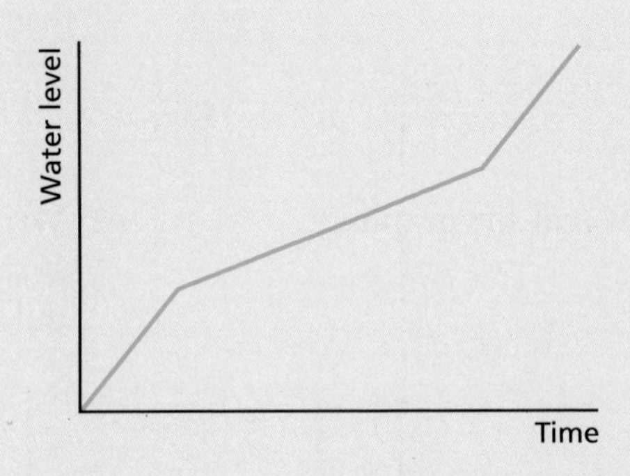

2 Scaling

This work will help you

- understand enlargement and scale factors
- use map scales
- work with ratios where different units are used, for example 5 cm : 2 m

A Spotting enlargements

A1 Decide whether shapes A to H are enlargements of this one. Explain for each shape.

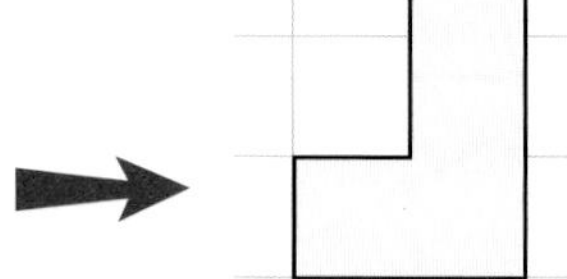

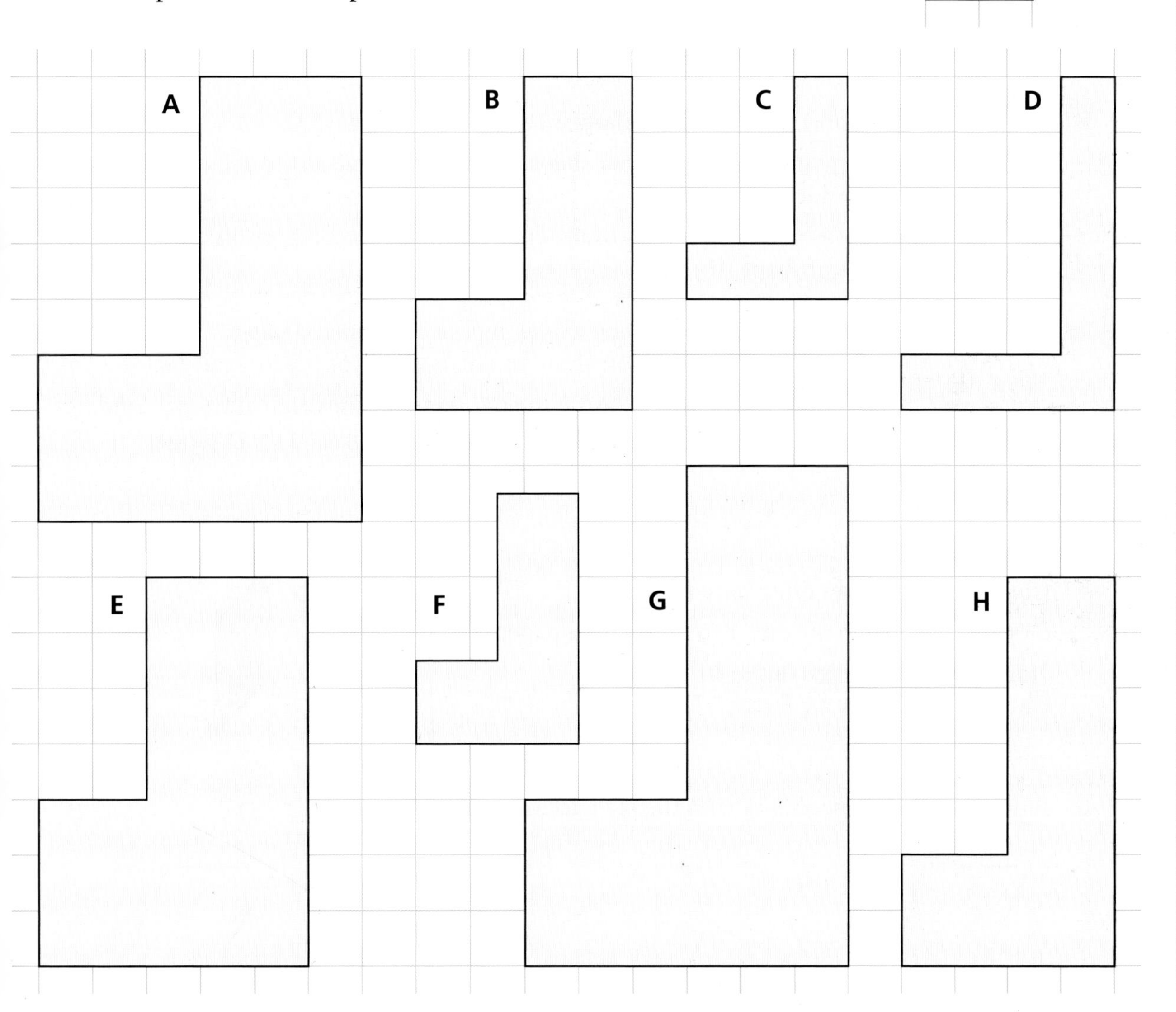

B Enlarging a shape

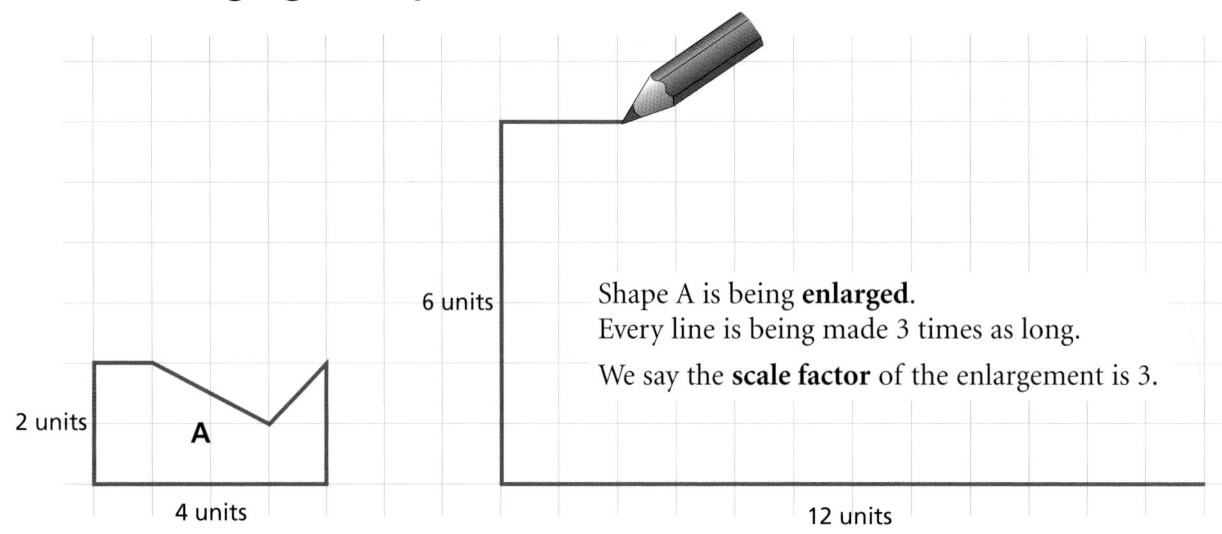

Shape A is being **enlarged**.
Every line is being made 3 times as long.

We say the **scale factor** of the enlargement is 3.

B1 (a) Copy shape A and complete the enlargement of it.
Check that every side of the enlarged shape is 3 times as long as the corresponding side of shape A.

(b) What can you say about the angles of the enlarged shape in comparison with the angles of shape A?

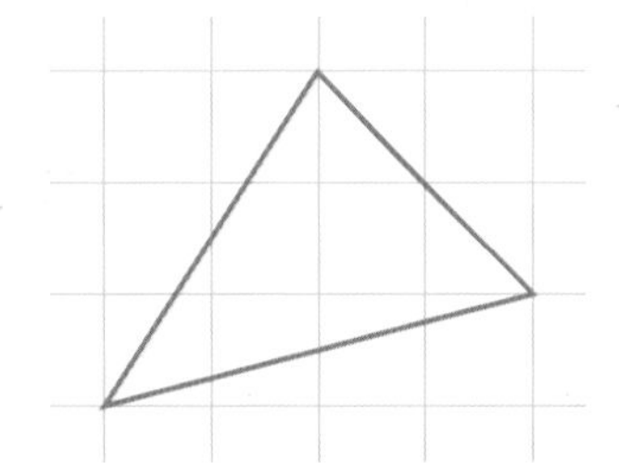

B2 (a) Copy this triangle on to squared paper.
Now draw an enlargement of it with scale factor 2.

(b) Measure the perimeter of the original triangle and that of the enlargement. How do they compare?

B3 Clubs and businesses often have a logo.
Design a small logo of your own on squared paper.
Now make an enlargement of it.
Choose your own scale factor.

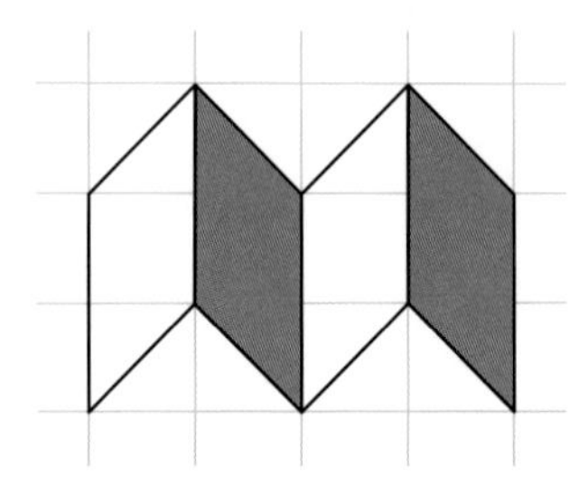

B4 Enlarge each of these shapes with the scale factor given.

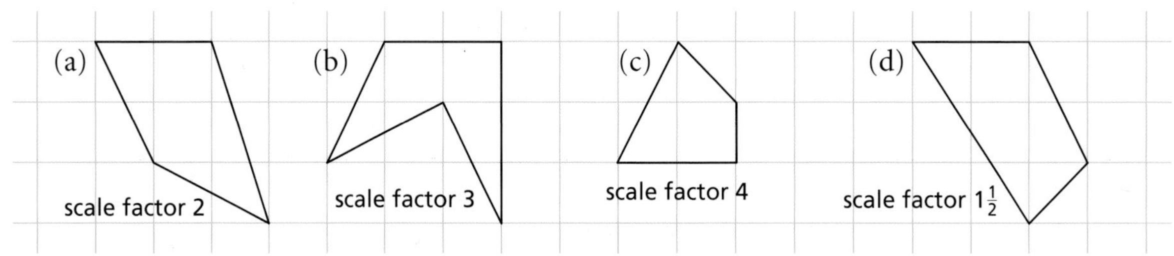

C Scaling down

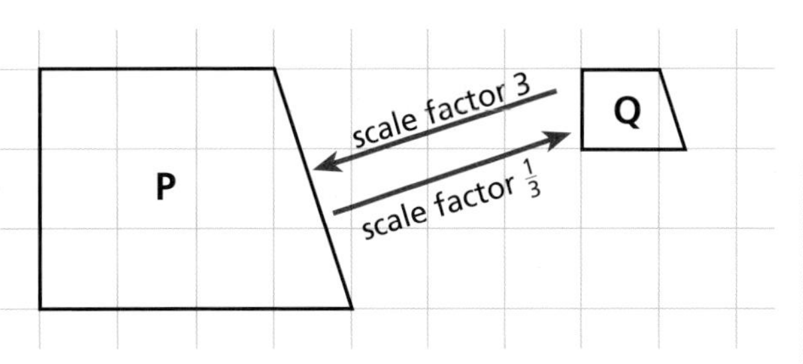

Shape P is an enlargement of Q with scale factor 3.

Although Q is smaller than P, we say
Q is an enlargement of P with scale factor $\frac{1}{3}$,
because every line in Q is $\frac{1}{3}$ as long as in P.

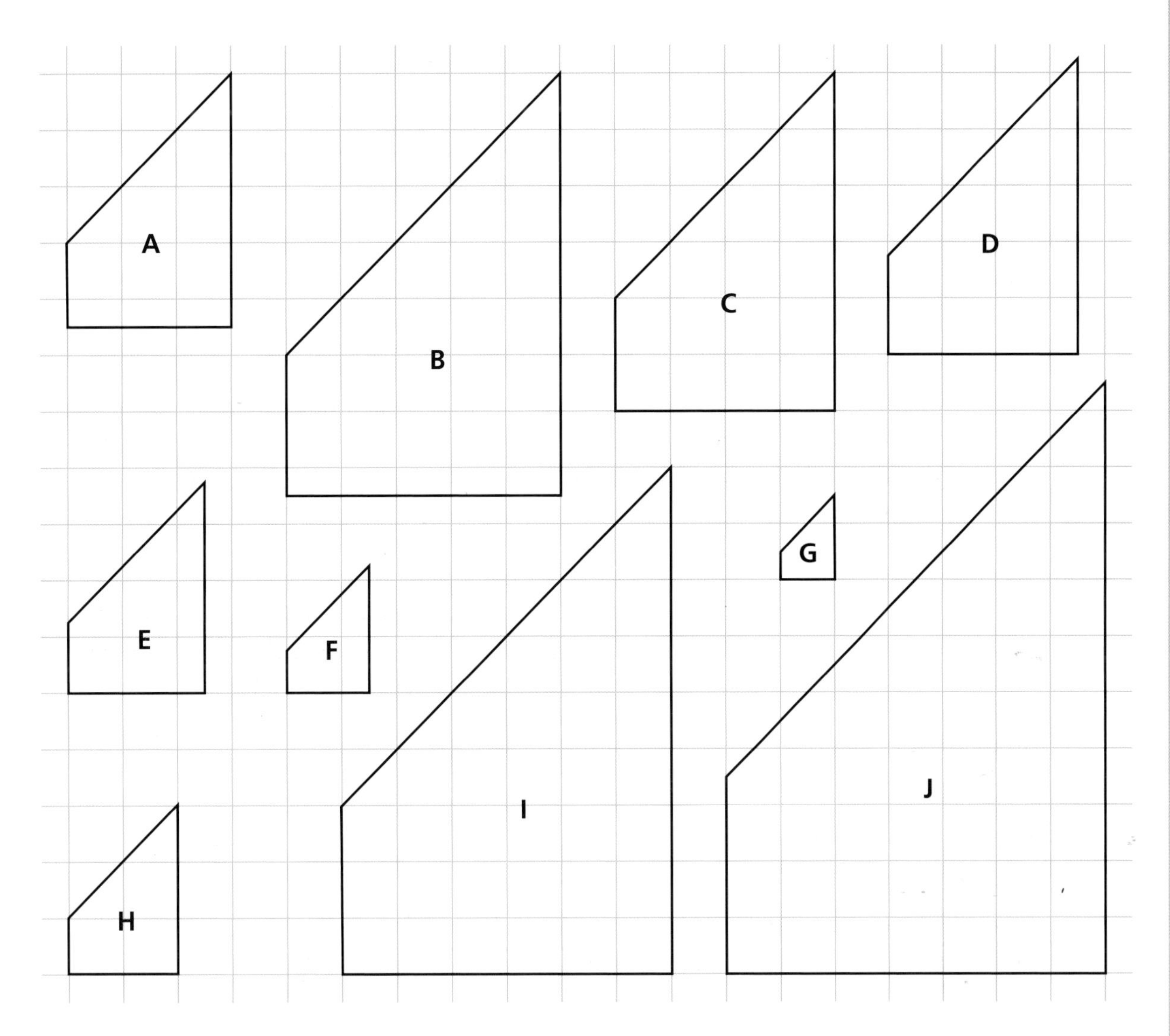

- Which shape is an enlargement of C with scale factor $\frac{1}{2}$?
- Which is an enlargement of I with scale factor $\frac{1}{3}$?

- What is the scale factor of the enlargement from B to G?
- What is the scale factor of the enlargement from H to A?
- What is the scale factor of the enlargement from A to H?

Make up other questions and answer them.

D Scales for maps and drawings

This is a map of a village.
You can think of it as a scaled-down drawing.

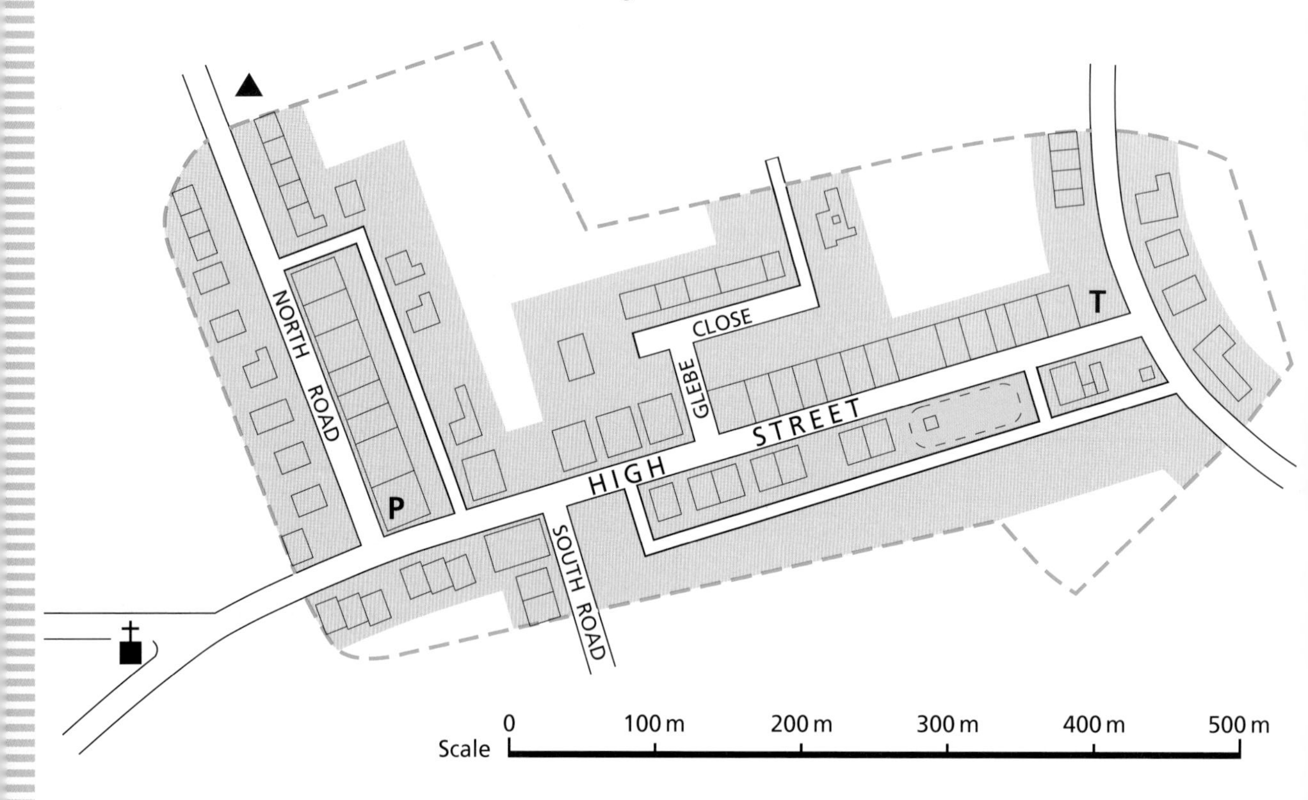

D1 Put a piece of string between the church (church symbol) and the post office (**P**). Mark the string at each place and then put the marks against the scale. Write down the distance in metres.

D2 Find the distance between each of these.

(a) The post office and the youth hostel (▲)

(b) The public telephone (**T**) and the post office

(c) The post office and the junction of South Road and High Street

(d) The public telephone and the junction of Glebe Close and High Street

D3 The dashed line shows a nature walk round the village. How long is the walk

(a) in metres (b) in kilometres

D4 What distance does 1 cm on the map represent?

This map shows the east end of the village, drawn to a larger scale.

D5 What does 1 cm on this map represent?

D6 (a) Find the dimensions of Holmcroft.

(b) The front of The Manor faces the road. How long is the front of the building?

(c) There is a tree behind Holmcroft. How far is its centre from the back of Holmcroft?

(d) How long is the pond at the back of The Manor?

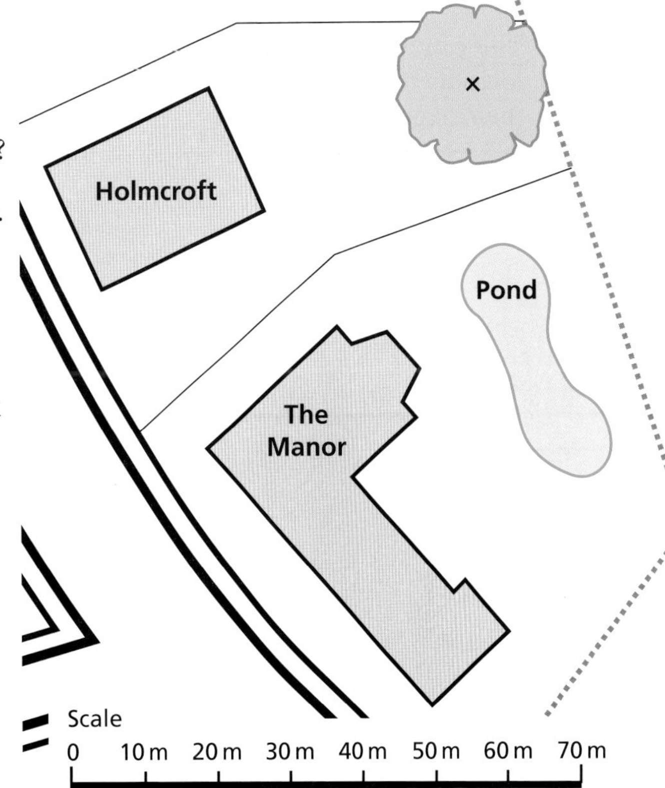

D7 This is a map of the Isle of Wight.

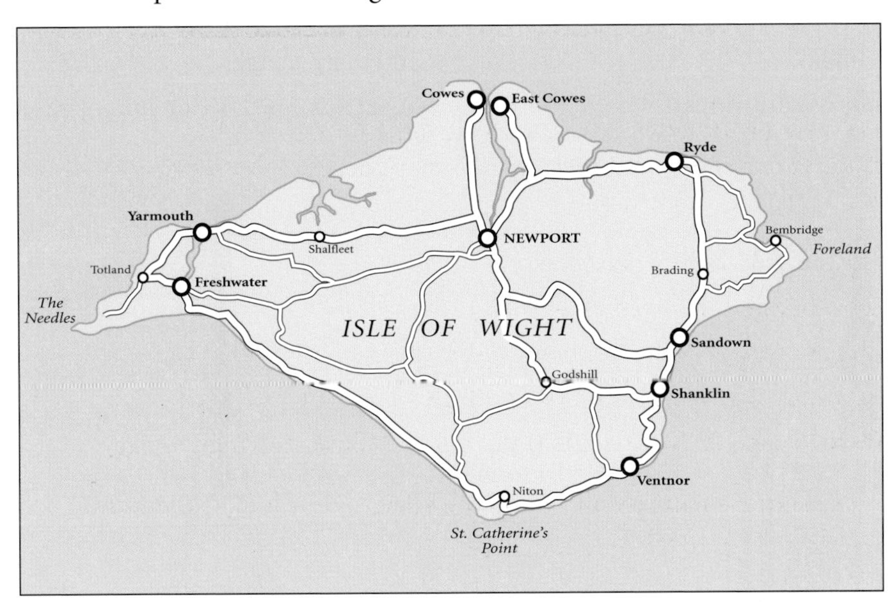

The real distance from The Needles to Foreland is 36 km.

(a) What does 1 cm on the map represent?

(b) Estimate the distance in km round the coast of the island.

D8 These buildings are drawn to a scale where 1 cm represents 50 metres. Measure them and work out their actual height.

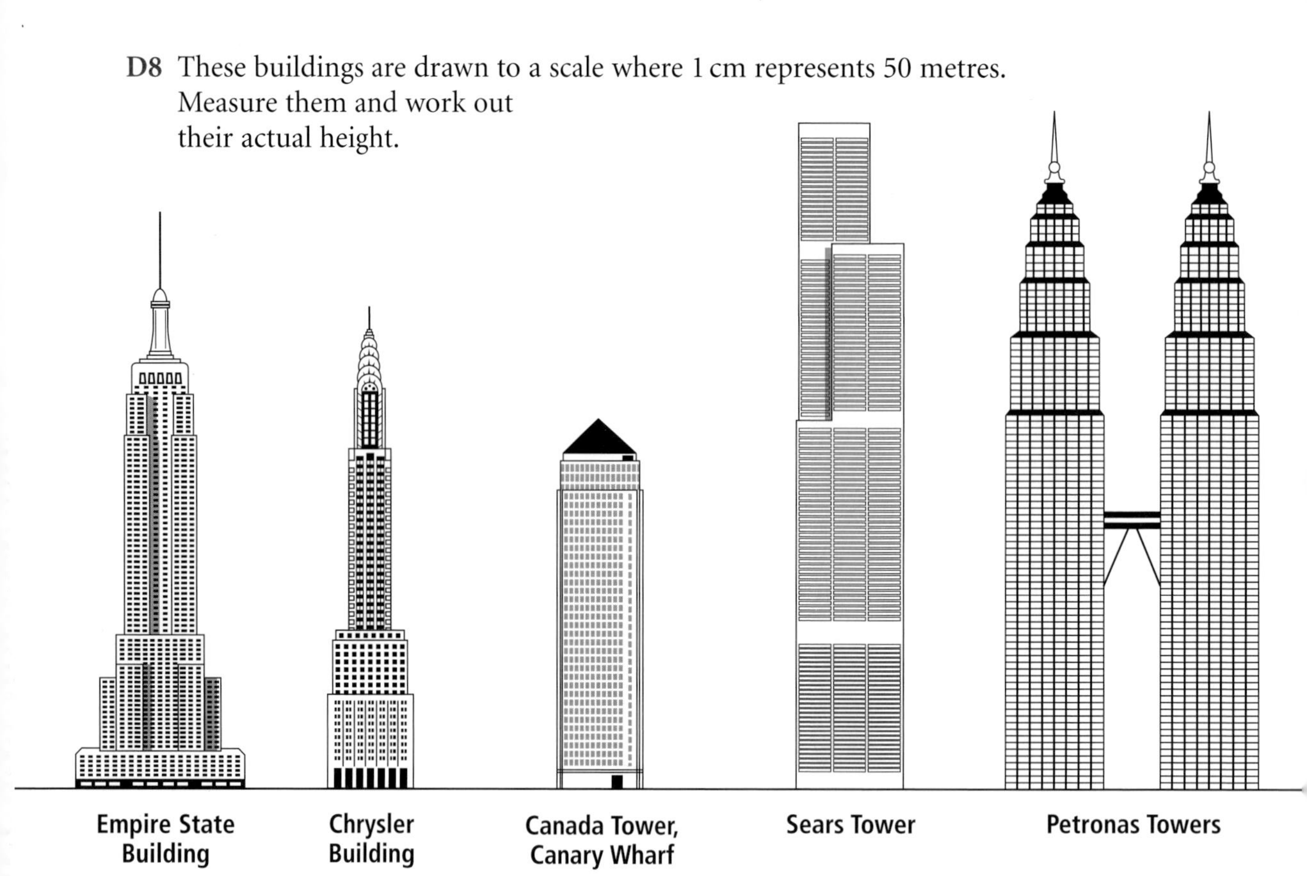

D9 The real Chartres Cathedral is 110 metres high.
What scale is this drawing of it?

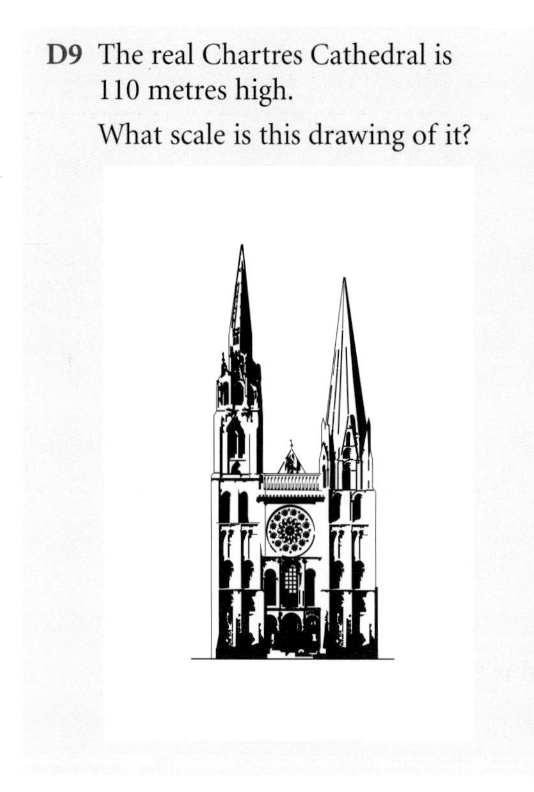

D10 The real Eiffel Tower is 300 metres high.
What scale is this drawing of it?

D11 This sketch shows the measurements of a rectangular field.

(a) Make an accurate scale drawing of the field using a scale of 2 cm to 1 metre.

(b) Find the length of the path shown as a dotted line.

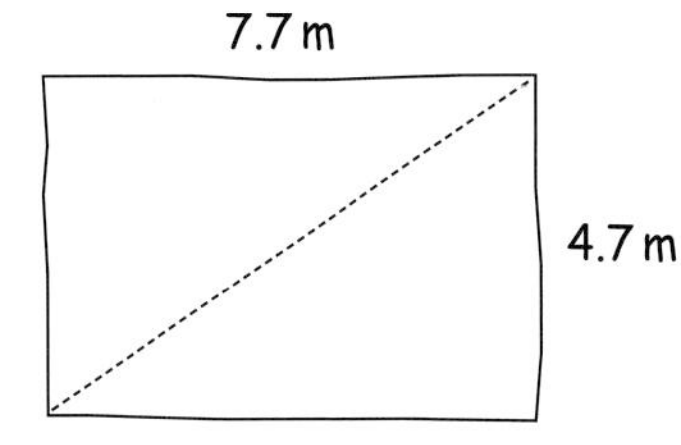

D12 What will these measurements of the Akashi Kaikyo bridge be, drawn to a scale where 1 cm represents 200 metres?

(a) The 1990 metre length of the central span

(b) The 300 metre height of the towers

D13 These four maps show lakes in the Lake District.
They are drawn to different scales.
Buttermere lake (marked B) is 2 kilometres long.

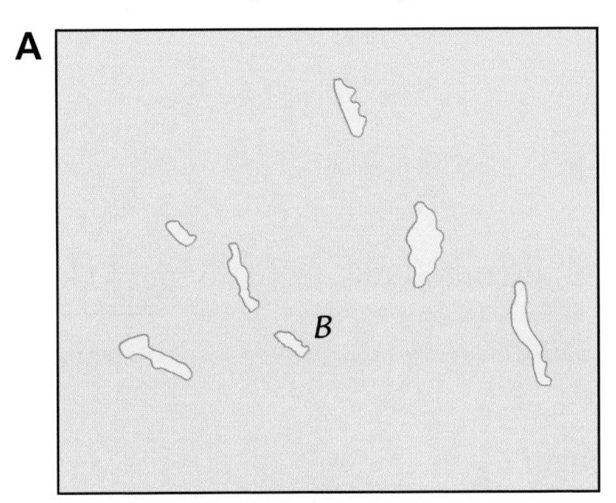

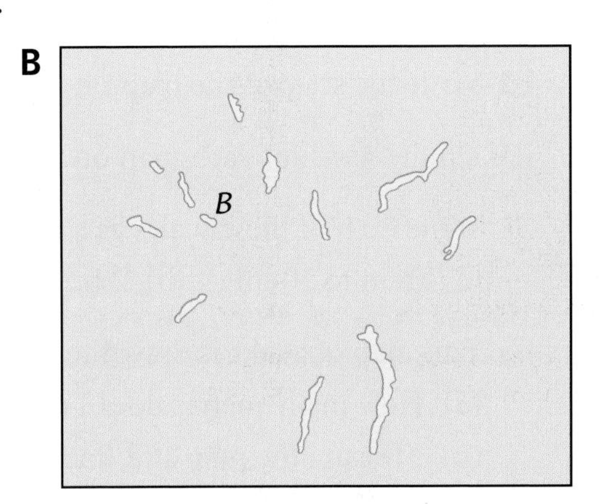

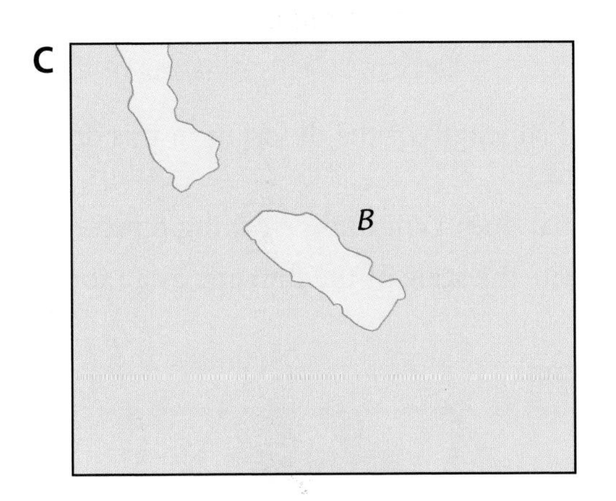

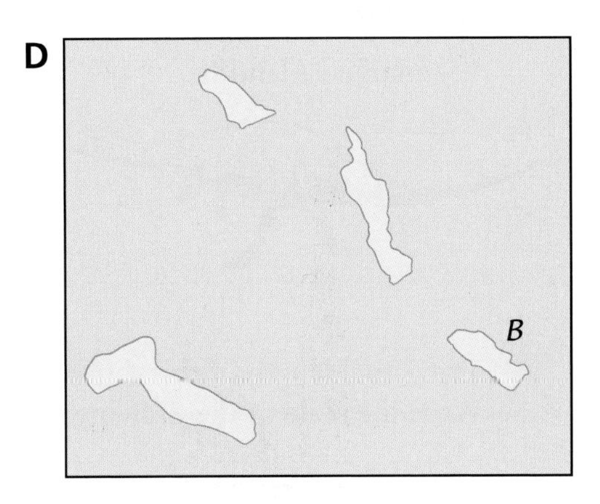

These are the scales for three of the maps.

1 cm represents 2 km.

1 cm represents 1 km.

1 cm represents 5 km.

(a) Which scale goes with which map?

(b) What is the scale of the remaining map?

E Ratios

Another way to give the scale of a map or plan is to write the **ratio**

distance on map or plan : actual distance

To do this we must measure both distances in the **same units**.

For example, on this room plan, 1 cm stands for 2 m.

So the scale of the plan, as a ratio, is **1 cm : 2 m**
Change the 2 m into centimetres: **1 cm : 200 cm**
Leave out the units and write it as **1 : 200**

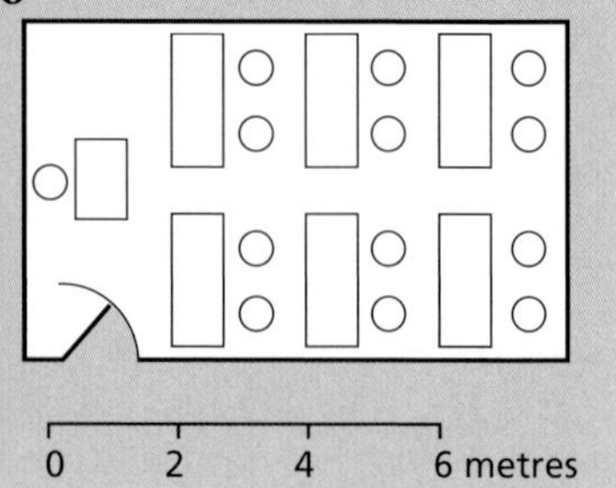

E1 Write each of these map scales as a ratio.

(a) 1 cm to 50 cm (b) 1 cm to 5 m (c) 1 cm to 200 m (d) 1 cm to 1 km

E2 Write the scale of the map on page 16 as a ratio.

E3 Do the same for each map on page 17.

E4 Write each of these scales as a ratio.

(a) 2 cm to 50 cm (b) 2 cm to 1 m (c) 5 cm to 1 m (d) 5 cm to 1 km

E5 This map of a lake is drawn to a scale of 1 : 2500.

(a) How many metres does 1 cm represent?

(b) Measure the map and find the actual length of the lake (the dotted line).

(c) Find the distance between the two places marked A and B.

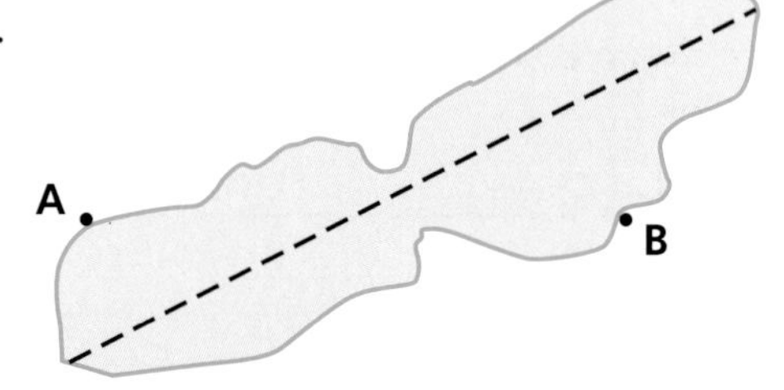

E6

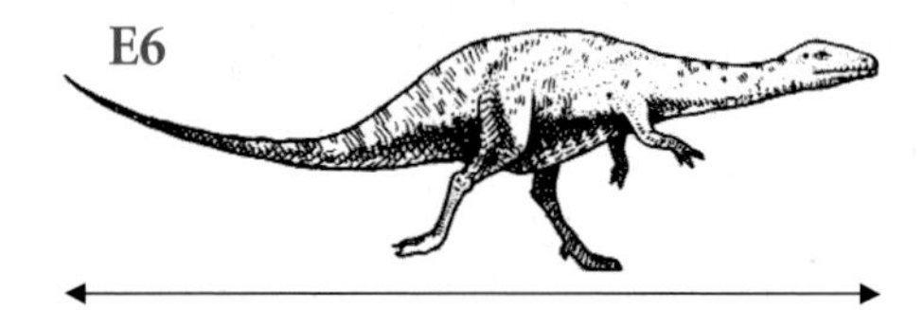

The actual length of the dinosaur in this drawing is 15 metres.

(a) What does 1 cm in the drawing represent?

(b) Write the scale of the drawing as a ratio.

E7 The actual length of this dinosaur is 30 metres.
Write the scale of the drawing as a ratio.

E8 A lizard is 20 cm long, weighs 160 g and has a stomach capacity of 4 millilitres.
An alligator is 1.2 m long, weighs 40 kg and has a stomach capacity of 1.2 litres.

Find these ratios in the form 1 : …

(a) length of lizard : length of alligator

(b) weight of lizard : weight of alligator

(c) stomach capacity of lizard : stomach capacity of alligator

What progress have you made?

Statement | **Evidence**

I can draw an enlargement with a given scale factor.

1 On cm squared paper, draw an enlargement of this shape with scale factor 3.

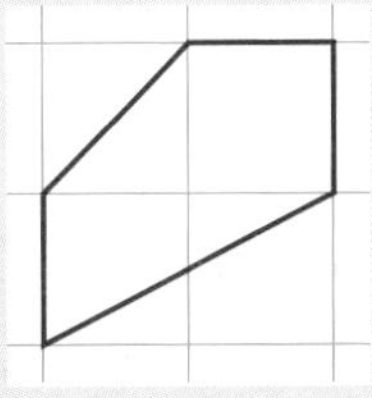

I can work out the scale factor of an enlargement.

2 What is the scale factor of the enlargement

(a) from P to Q (b) from Q to P

(c) from R to P (d) from Q to R

(e) from R to Q

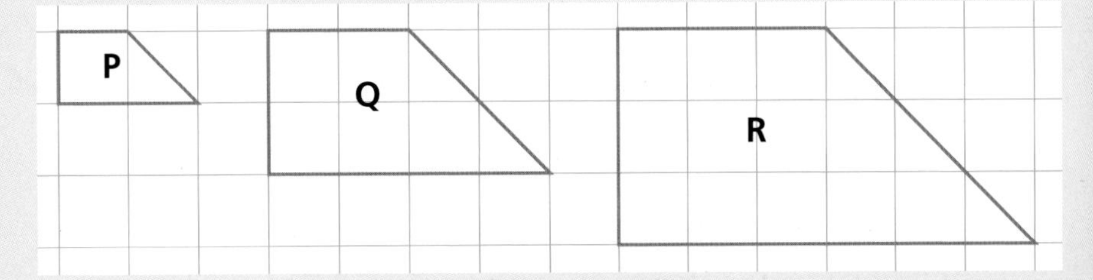

I can find a distance from a scale drawing.

3 This is a plan of a cinema.

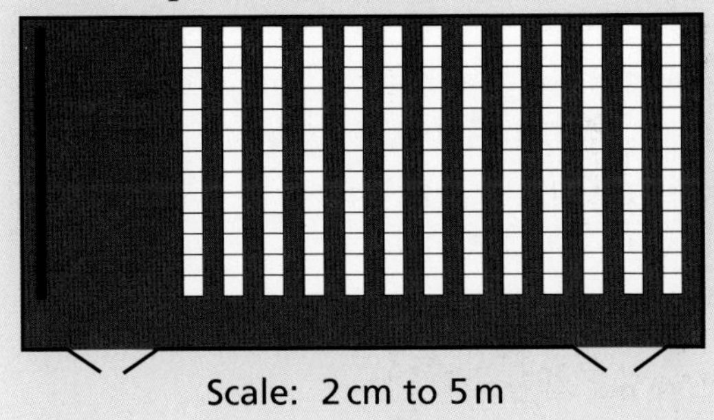

Scale: 2 cm to 5 m

Find the length and width of the real cinema.

I can work out lengths for a scale drawing.

4 Another cinema is 24 m long and 17.5 m wide.

(a) If you want to draw a plan of it to a scale of 1 cm to 5 m, how long and how wide should your plan be?

(b) How long and how wide will the plan be if it is drawn to a scale of 5 cm to 1 m?

I can find the ratio of two quantities expressed in different units.

5 Write each of these map scales as a ratio.

(a) 1 cm to 2 m (b) 2 cm to 5 m

(c) 1 cm to 50 m (d) 2 cm to 1 km

3 Graphs and charts

This work will help you
- read and interpret graphs and charts
- draw frequency charts and line graphs

A Off the record

This graph shows how many millions of CDs, cassettes and vinyl LPs were supplied to shops between 1988 and 1998.

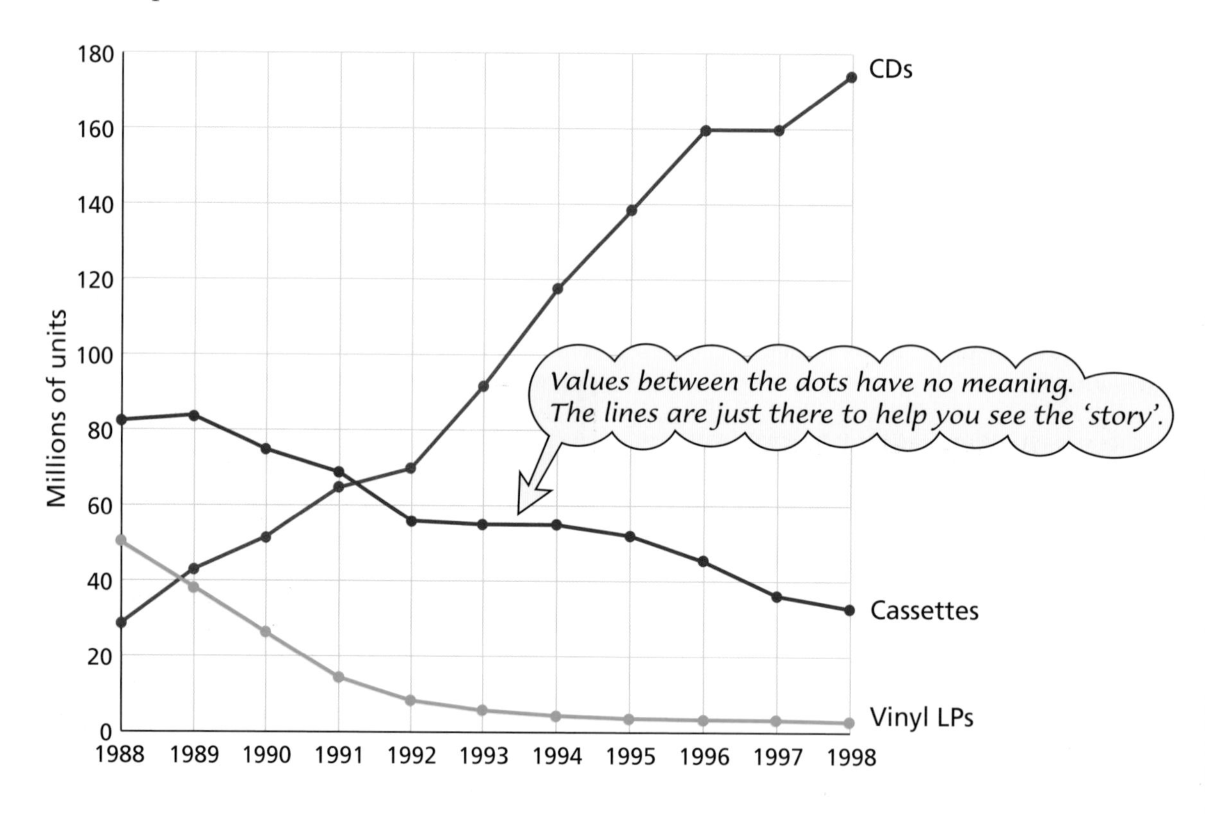

A1 Between 1988 and 1991, sales of vinyl LPs fell rapidly.
What happened to them after that?

A2 Describe briefly what happened to the sales of cassettes between 1988 and 1998.

A3 What was the most popular way of buying recorded music in (a) 1988 (b) 1994

A4 In which year did the number of CDs supplied overtake

(a) the number of LPs (b) the number of cassettes

B Drawing graphs and charts

Graphs

Average daily maximum temperature in °C in London (UK) and Auckland (NZ)

	Jan	Feb	Mar	Apr	May	Jun	Jul	Aug	Sep	Oct	Nov	Dec
London	7	7	10	13	16	20	22	21	19	15	10	8
Auckland	24	23	22	21	17	15	14	16	16	18	19	21

Temperature of oven in °C at one-minute intervals from switching on

Time in minutes	0	1	2	3	4	5	6	7	8
Temperature in °C	20	68	129	157	173	184	192	196	198

Frequency bar charts

Pupils' test scores

37	41	28	60	56	39	17	39	73	64
58	25	44	66	34	32	78	35	46	76
18	39	56	38	75	53	49	55	38	47
53	86	34	64	26	36	22	18	73	91
64	86	15	27	53	72	76	52	57	28
74	24	46	39	74	53	33	70	28	30

Times taken (in seconds) to run 100 metres

21.31	19.64	17.56	17.84	20.48	16.91	22.51	23.03
19.23	18.17	17.36	26.15	19.35	24.08	21.37	18.63
17.52	22.23	25.11	18.85	21.47	24.68	18.99	23.90
27.01	19.30	17.73	20.05	25.16	20.43	17.39	16.50
17.73	23.23	18.01	24.57	26.18	19.93	18.44	23.74

C Which graph?

Here are the midday temperatures (in °C) at two places, over a period of 20 days.

Day	1	2	3	4	5	6	7	8	9	10	11	12	13	14	15	16	17	18	19	20
Sanpam	10	11	13	13	14	15	16	17	16	18	19	20	19	20	21	22	23	22	23	24
Tolero	23	21	20	18	19	19	19	18	17	17	15	16	15	14	14	15	14	13	12	12

This information is shown in each of the graphs on the opposite page.

Go through questions C1 to C7 below, one by one, and say

(a) which graph is easiest to answer the question from

(b) if there are any graphs from which you cannot answer the question

C1 Which place had the highest midday temperature during the period?

C2 At which place was the temperature most often 22°C or above?

C3 At which place was the range of temperatures greater?

C4 Which place was hotter, overall?

C5 Was the temperature fairly steady at each place, or did it go broadly up during the period or broadly down?

C6 At which place was the temperature most often in the range 14–21°C?

C7 What was the most common temperature range, 10–13°C, 14–17°C, 18–21°C or 22–25°C, at each place?

C8 The time that is halfway between midday on day 1 and midday on day 2 is midnight.

On graph A, is it sensible to read off midnight temperatures halfway between the marked points? If not, why not?

Graph A

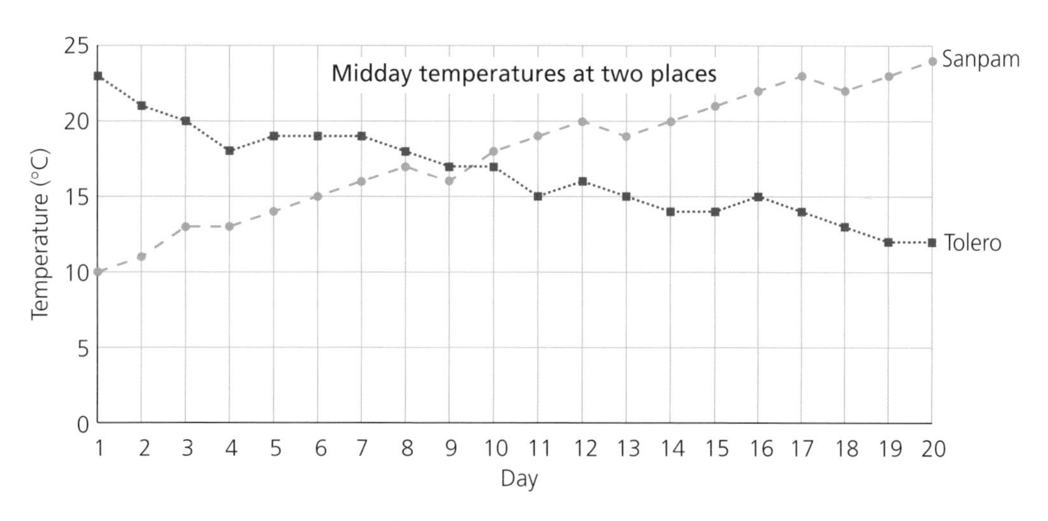

Graph B

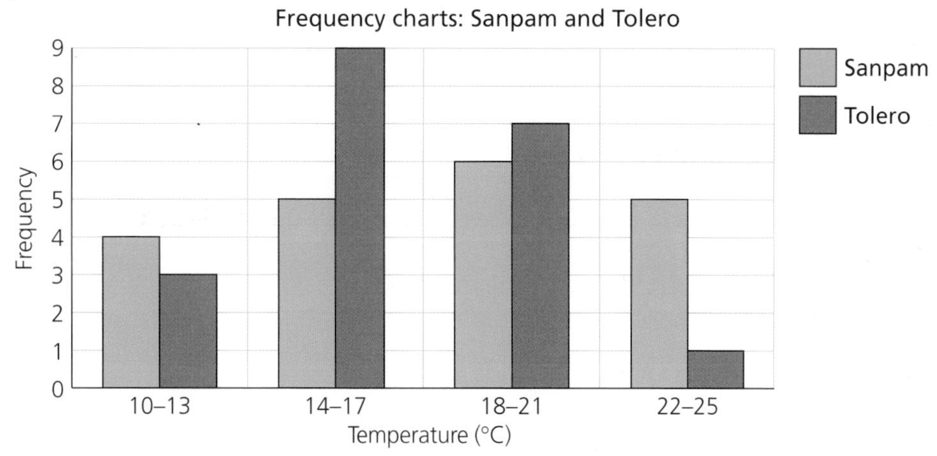

Graph C

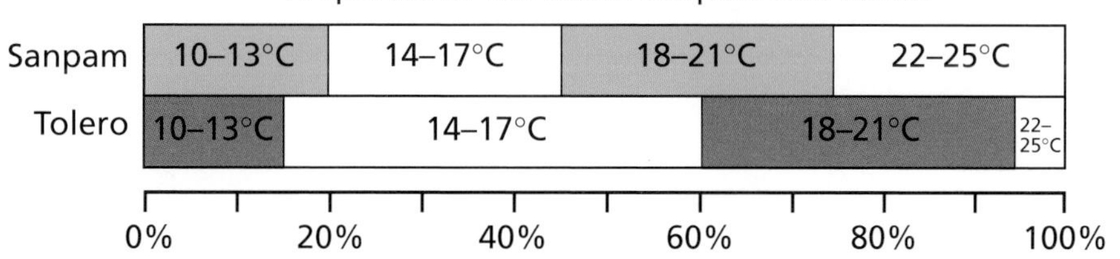

D Road accidents

All these charts refer to road accidents in Cornwall in 1999.

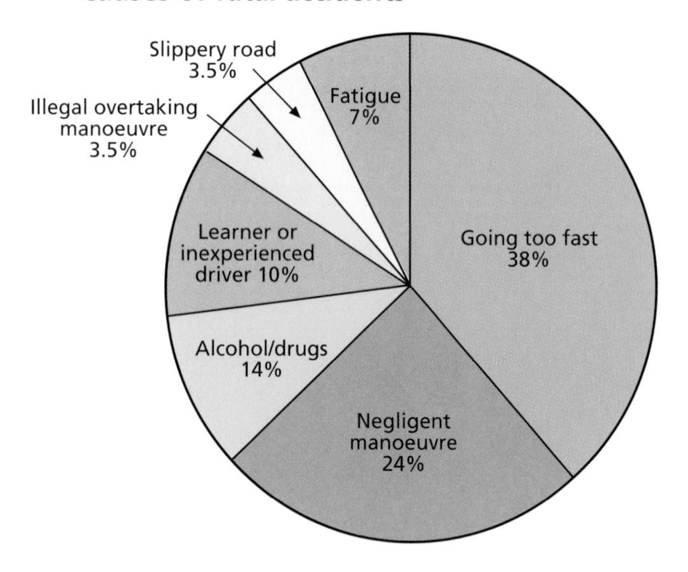

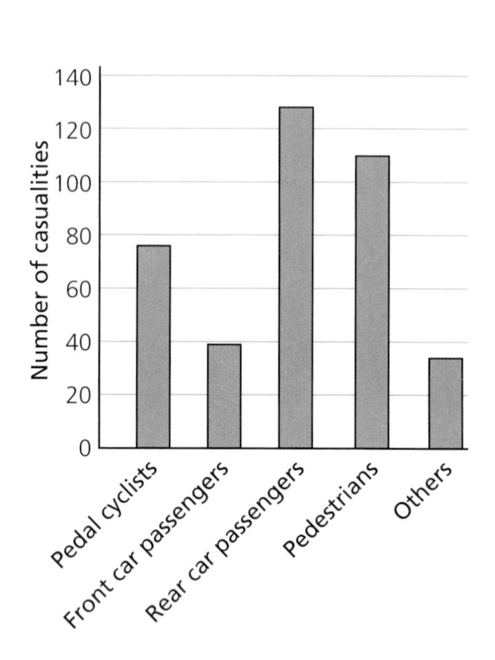

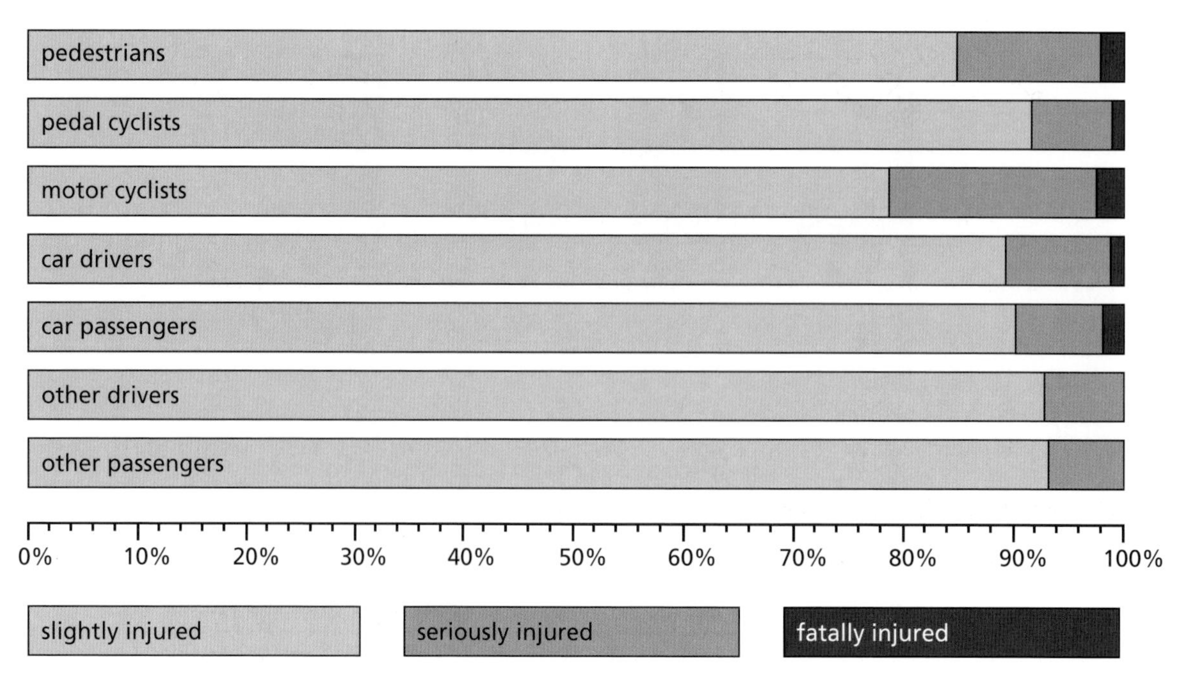

Information on these two pages is taken from the leaflet 'Road accidents in Cornwall', published in 2000 by the Cornwall County Council Road Safety Unit.

Age groups of casualties: pedestrians, cyclists and motor cyclists

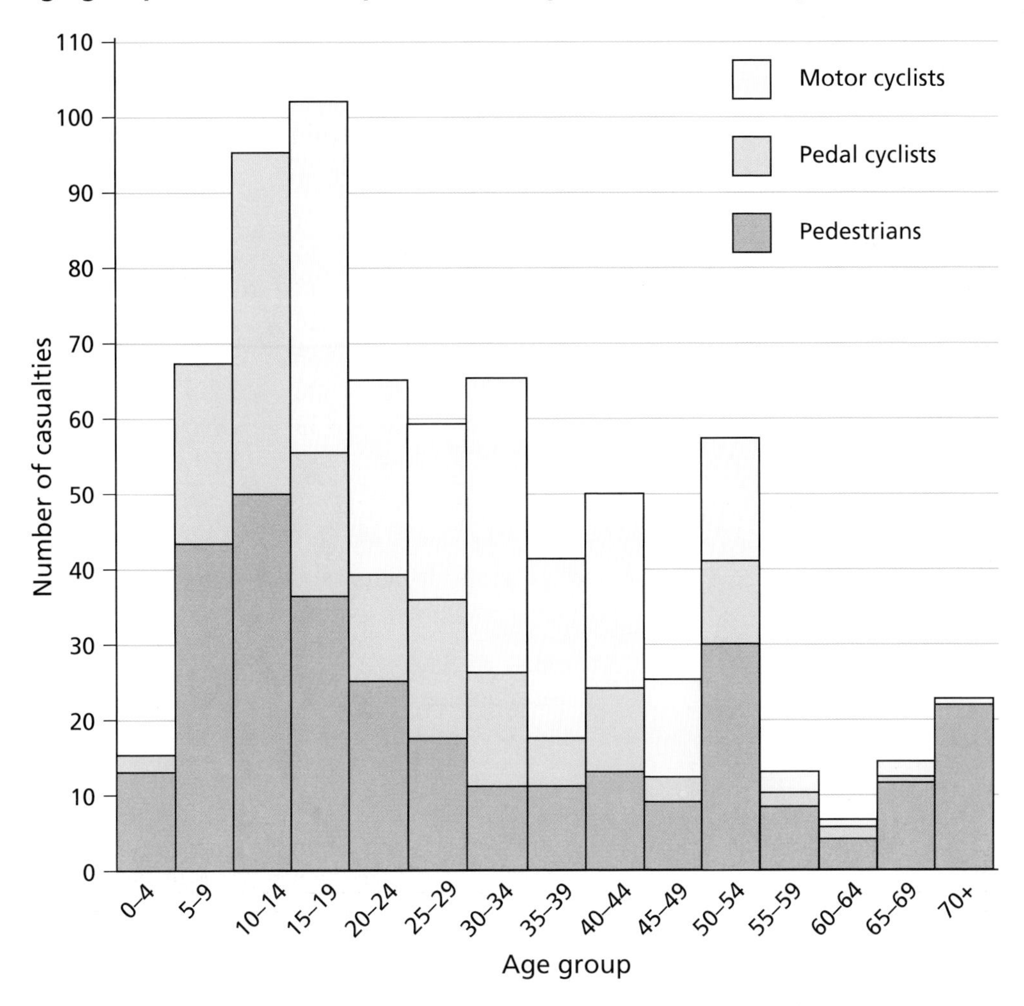

Use one or more of the charts to answer these questions.

D1 What percentage of pedestrian casualties suffered serious or fatal injuries?

D2 How many pedestrian casualties were in the 0–14 age group?

D3 (a) What were the two most common causes of fatal accidents?

(b) What percentage of fatal accidents do these two causes together account for?

D4 What was the most common age group for

(a) pedestrian casualties (b) motorcycle casualties (c) pedal cycle casualties

***D5** Does the information given here support the view that it is more dangerous for children to sit in the rear than in the front of a car? Explain your answer.

D6 Use these charts to prepare three questions and give them to someone else to answer. Make sure you know the answers yourself!

What progress have you made?

Statement	Evidence
I can draw a frequency bar chart, choosing the groupings myself.	**1** Draw a grouped frequency bar chart for these pupils' test marks.

34	17	23	22	20
18	42	16	11	28
41	35	44	23	19
15	40	17	26	29
13	46	25	36	33
27	36	32	36	38
24	30	36	33	28
31	17	14	23	24

Statement	Evidence
I can draw a line graph, choosing the scales myself.	**2** Draw a graph to illustrate this data, which shows how a plant grew.

Number of days after planting	Height in cm
0	3.4
1	5.0
2	8.5
3	12.7
4	15.6
5	17.2

Statement	Evidence
I can use a number of graphs and charts to find the answers to questions.	**3** Use the graphs on the previous two pages to answer these questions.

(a) For what type of vehicle was a casualty most likely to be a serious one?

(b) How many pedal cycle casualties were there in the 0–14 age group?

(c) In which age groups were there more pedestrian casualties than motorcycle casualties?

4 Solving equations

This is about solving algebraic equations.
The work will help you

- form an equation yourself
- solve equations using algebra

A Balances review

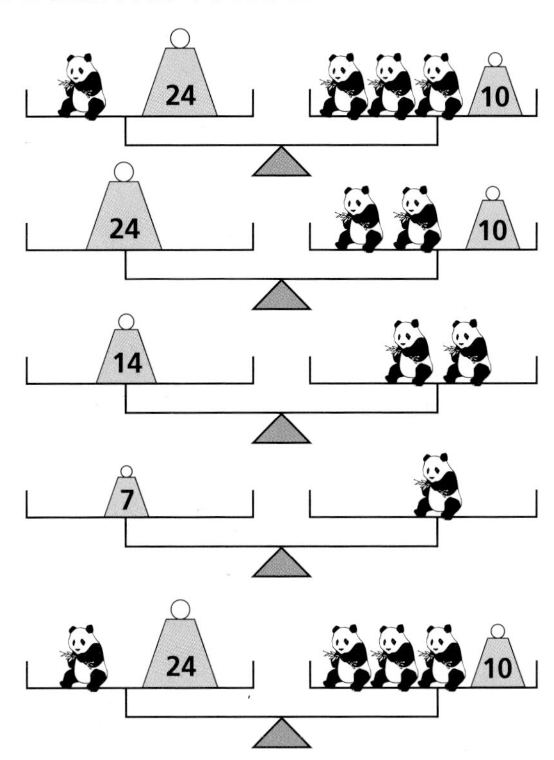

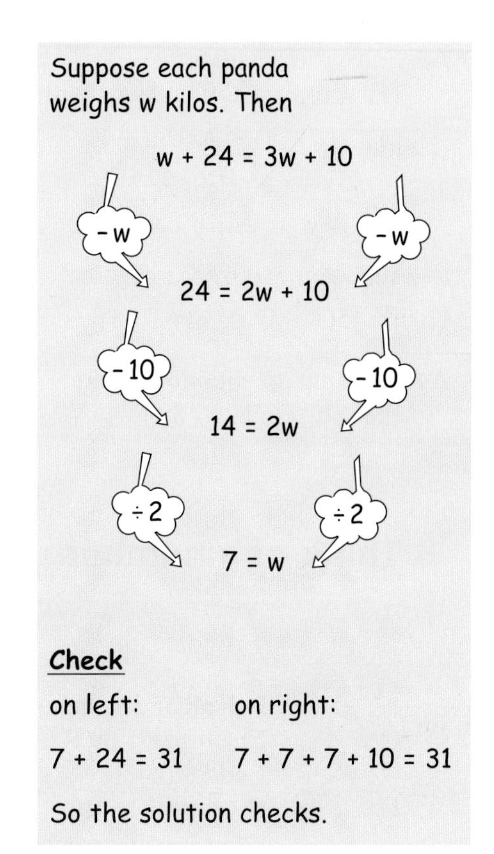

Suppose each panda weighs w kilos. Then
w + 24 = 3w + 10
- w
- w
24 = 2w + 10
- 10
- 10
14 = 2w
÷ 2
÷ 2
7 = w
Check
on left:
7 + 24 = 31
on right:
7 + 7 + 7 + 10 = 31
So the solution checks.

A1 Solve each of these balance puzzles.
Show your working and check your answers.

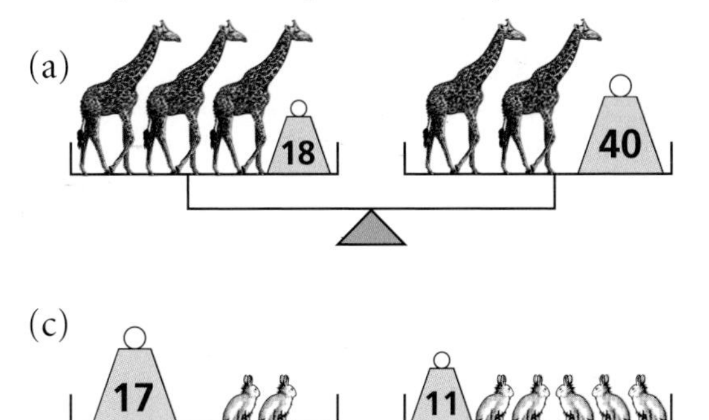

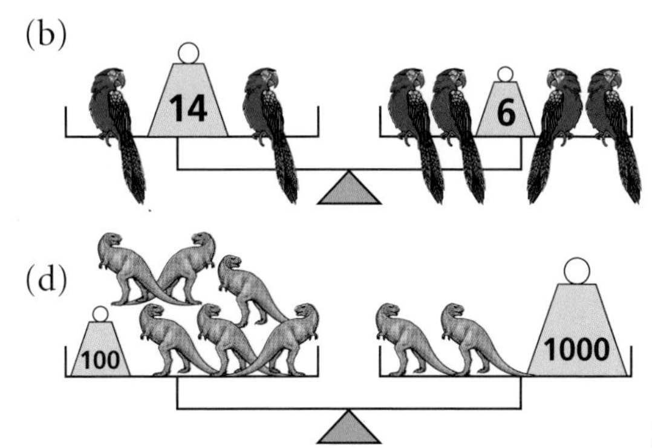

A2 Solve each of these equations.
Show your working and your check.

(a) $5t + 7 = 27$
(b) $7 + 2n = 22$
(c) $5d + 8 = 3d + 28$
(d) $4x + 11 = 3x + 16$
(e) $5q + 11 = 24 + 3q$
(f) $17 + 4n = 2n + 37$
(g) $2.5 + 3m = 4 + m$
(h) $2n + 37 = 1 + 12n$
(i) $2e + 12 + e = 5e + 9$
(j) $0.5a + 12 = 3a + 7$
(k) $0.1h + 2 = h + 0.2$
(l) $\frac{x}{2} + 22 = x + 1$

A3 Solve these equations.
(First replace the brackets with equivalent expressions.)

(a) $2(d + 4) + 2 = 5d + 1$
(b) $3(b + 4) + 1 = b + 25$
(c) $3x + 62 = 10(x + 2)$
(d) $3(k + 6) = 5(k + 2)$
(e) $2g + 35 = 4(g + 7)$
(f) $7(h + 6) = 11(h + 2)$
(g) $24 = 3(k + 5)$
(h) $5 + 3(m + 2) = 4 + 2(m + 5)$
(i) $5(y + 17) = 6(y + 14)$
(j) $0.5(h + 24) = 2h + 6$

A4 Make up an equation where the solution is $t = 5$.

B Think of a number

I think of a number.
I multiply it by 8.
I add on 6.
My answer is 110.
What was my number?

I think of a number.
I multiply it by 7.
I add on 9.
My answer is 13 times the number I started with.
What was my number?

B1 Turn this number puzzle into an equation.

Solve the equation and check your solution works in the puzzle.

I think of a number.
I multiply it by 6.
I add on 20.
My answer is 10 times the number I started with.
What was my number?

B2 Turn each of these number puzzles into equations and solve them.

(a) *Jo thinks of a number.*
She multiplies it by 9.
She adds on 18.
Her answer is 11 times the number she started with.

(b) *Harry thinks of a number.*
He multiplies it by 4.
He adds on 100.
His answer is 8 times the number he started with.

B3 Copy and complete the working for this puzzle.

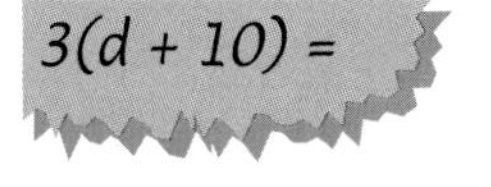

Damien thinks of a number.
He adds on 10, and then multiplies by 3.
His answer is 42.

What number did he start with?

B4 Solve this puzzle.

Jay thinks of a number.
He adds 4 to his number, then multiplies by 5.
He adds on 10. His answer is 7 times what he started with.

B5 Solve these two puzzles.

(a) *Hamish thinks of a number.*
He adds on 12.
Now he multiplies by 4.
He adds on 2.
His answer is 9 times as big as his starting number.

(b) *Anne thinks of a number.*
She adds on 1.
She multiplies by 7.
She adds on 3.
Her answer is 9 times the number she started with.

B6 Adie and Biff are both doing 'think of a number' puzzles.
They both start with the same number.
We will use n for the number they both start with.

Adie multiplies the number by 5 and then adds on 1.

Biff multiplies the number by 3 and then adds on 9.

They are surprised to find that they both end up with the same number.
So $5n + 1 = 3n + 9$

Solve the equation to find what number they both started with.
Check your solution works.

B7 Sofima and Will both think of the same number.
Sofima multiplies her number by 5 and adds 10.
Will multiplies his number by 3 and adds 28.
They both get the same answer.

What number were they thinking of?

B8 Cath and Janet start with the same number.
Cath just adds 29 to her number.
Janet multiplies her number by 7 and adds 5.
They get the same answer.

What was their starting number?

B9 John and Majid start with the same number.
John first adds 2 to his number and then multiplies by 5.
Majid just adds 22 to his number.
Their answers are equal.

What number did they each start with?

***B10** Today is Kirsty's birthday. In 50 years' time, Kirsty will be three times as old as she will be in 6 years' time.

How old is Kirsty today?

C Both sides

T

1 Discuss how to solve this problem.

Darren and Julie both think of the same number.
Darren multiplies his number by 5 and then takes off 16.
Julie just adds 4 to her number.
They both get the same answer.
What number were they thinking of?

2 Discuss how to solve these equations. $5m - 6 = 3m + 18$ $a - 6 = 3a - 26$

C1 Copy and complete this working to solve $s + 1 = 3s - 17$.
(You don't need to copy the 'bubbles'.)

Check your solution works.

$s + 1 = 3s - 17$

+ 17 | + 17

$s + \ldots\ldots = 3s$

– s | – s

$\ldots\ldots = \ldots\ldots$

÷ 2 | ÷ 2

$\ldots\ldots = \ldots\ldots$

C2 Aisha has solved this equation correctly.

Copy her solution and fill in the bubbles to show how she got from one line to the next.

$9z - 7 = 4z + 3$

– 4z | – 4z

$5z - 7 = 3$

.... | ....

$5z = 10$

.... | ....

$z = 2$

C3 Here is Martina's homework.
Correct it for her.
Write a short explanation to tell her what she has done wrong in each one.

(a)
$$4d - 2 = 2d + 6$$
$$4d = 2d + 4$$
$$2d = 4$$
$$\underline{d = 2} \times$$

(b)
$$3s - 2 = 5s - 8$$
$$2 = 2s - 8$$
$$10 = 2s$$
$$\underline{s = 5} \times$$

(c)
$$5h - 10 = 4h$$
$$5h - 14 = h$$
$$5h = h + 14$$
$$4h = 14$$
$$\underline{h = 3.5} \times$$

C4 Solve these equations.
Show your working and check your answers.

(a) $5n - 3 = 3n + 9$
(b) $b - 2 = 5b - 18$
(c) $2f + 11 = 4f - 5$
(d) $4w - 9 = 3w + 2$
(e) $7y - 12 = 4y - 3$
(f) $s + 2 = 3s - 5$
(g) $u - 20 = 6u - 120$
(h) $5w + 21 = 8w - 30$
(i) $6 + 6y = 8y - 10$
(j) $6g - 2 = 1 + 4g$
(k) $\frac{3}{4}p + \frac{3}{4} = p - 4$
(l) $0.6g - 1 = 0.1g - 0.5$
(m) $0.1f + 1 = 0.01f + 10$
(n) $\frac{a}{2} + 12 = a - 2$

C5 Solve these equations.
Where there are brackets, replace them with an equivalent expression first.

(a) $4(a - 3) = a + 30$
(b) $2x + 13 = 3(x - 1)$
(c) $4(y - 2) = 3(y + 1)$
(d) $2(b - 1) = 3b - 11$
(e) $7j + 12 = 3(j + 20)$
(f) $5t - 20 = 3(t + 1)$
(g) $18(x - 1) = 17(x + 1)$
(h) $24 = 4(b - 1)$
(i) $0.5(j + 4) = j - 1$
(j) $2(b - 4) + 9 = 4(b - 2)$
(k) $4 + 2(t - 6) = t$
(l) $4(k + 1) - k = 8k - 20$
(m) $s - 4 = 0.4(s - 1)$
(n) $4z + 2 = 2(z - 1)$

C6 Mary and Tracey both think of the same number.
Mary multiplies the number by 6 and then takes off 16.
Tracey multiplies the number by 5 and then takes off 4.
They both get the same answer.

What number were they thinking of?

C7 Jenny and Bob are twins.
Jenny takes her age, multiplies it by 6 and takes off 15.
Bob takes his age, multiplies by 2 and adds 53.
The answer they both get is the age of their Uncle Fred.

Work out the ages of Uncle Fred and the twins.

C8 Andy, Barrie and Chris are brothers.
Chris is three times as old as Andy.
Barrie is 20 years older than Andy, but 4 years younger than Chris.

How old is each brother?

C9 Peter looks at today's date.
He doubles it and takes away 17.
His answer is the number of days this month.
He gets the same answer if he adds 6 to today's date.

What month is it?

C10 Jason has these four cards. $4w - 1$ | $6w - 19$ | $3w + 11$ | $2w + 19$

(a) He picks two of the cards. $4w - 1$ | $2w + 19$
Find w so that both these cards work out to the same value. What is the value?

(b) Look at the pair of cards left over: $6w - 19$ | $3w + 11$

Find w so that both these cards have the same value. What is the value?

C11 Pick a different pair of cards from Jason's four.

(a) Find a value of w that makes these cards have the same value.

(b) Do the same with the pair of cards left over.

C12 Repeat question C11 for yet another way of pairing the cards.

*__C13__ There is a value for x that makes all these cards have the same value.

$3x + 15$ | $5x + 3$ | $x +$ | $2x +$ | $x - 9$

(a) Work out the value of x.

(b) Write out the expression on each torn card.

What progress have you made?

Statement	Evidence
I can solve equations involving addition, with letters on both sides.	**1** Solve these equations. Show your working and your check. (a) $6y + 2 = 2y + 30$ (b) $f + 32 = 6f + 7$ (c) $0.5t + 1.6 = t + 0.1$
I can solve equations involving subtraction, with letters on both sides.	**2** Solve these equations. Show your working and your check. (a) $h + 1 = 4h - 20$ (b) $4r + 1 = 10r - 17$ (c) $s - 10 = 3s - 18$
I can solve equations involving brackets.	**3** Solve these equations. Show your working and your check. (a) $2(g - 1) = 5g - 17$ (b) $x + 17 = 4(x - 1)$ (c) $4(y + 3) = 2(y + 10)$
I can form equations from number puzzles and solve them.	**4** Solve these number puzzles. Show all your working clearly. (a) John thinks of a number. He multiplies it by 3 and adds 24. His answer is 5 times as big as his starting number. Find John's original number. (b) Viv thinks of a number. She takes off 6 and then multiplies by 2. Her answer is 20 more than her starting number. What number did she think of? (c) Pat and Jim think of the same number. Pat doubles her number and adds 1. Jim takes 1 off his number and then multiplies by 4. They both get the same answer. What number did they both think of?

5 Solids

This work will help you
- identify cross-sections of solids
- draw cross-sections
- identify the planes of symmetry of a solid

T

A Cross-sections

A screwdriver is being lowered into some water.

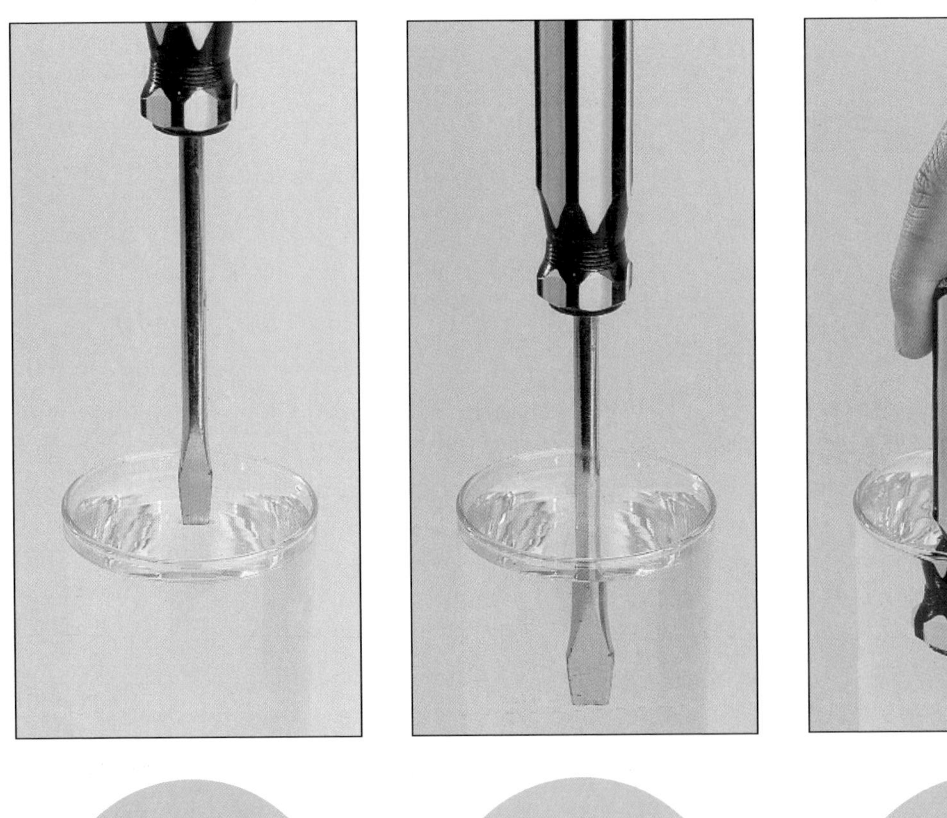

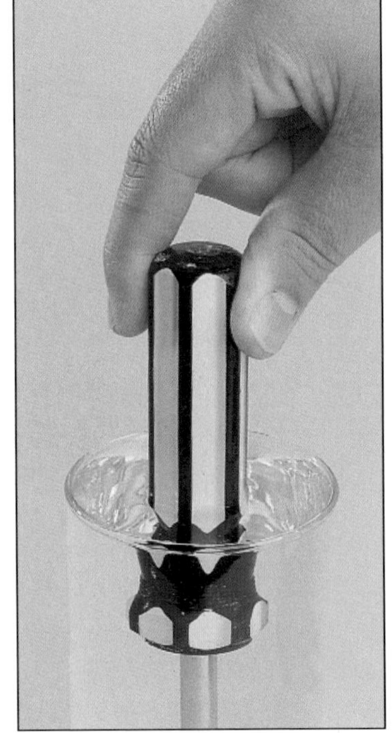

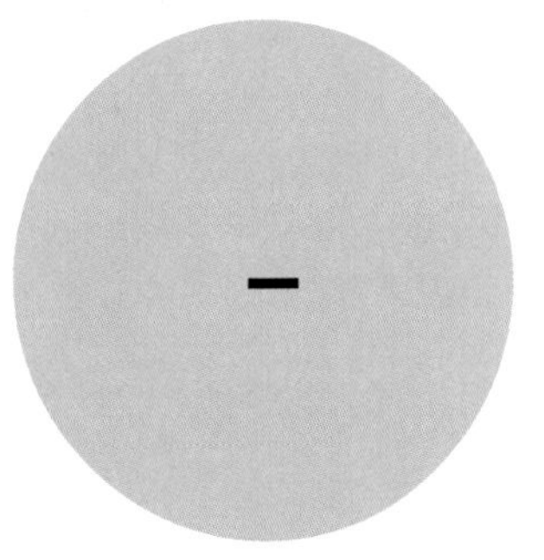

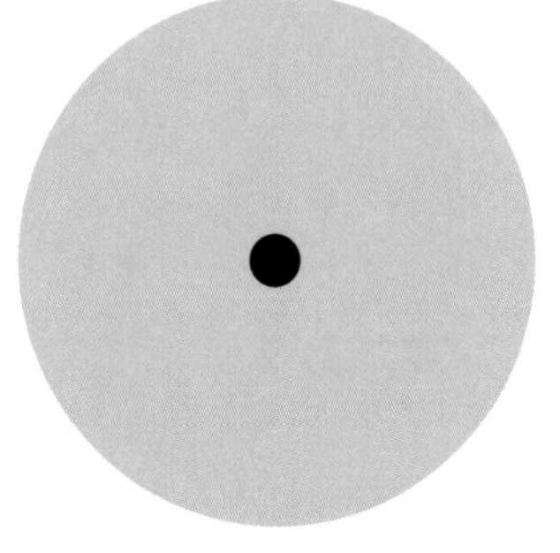

The drawing for each photo shows the cross-section of the screwdriver at water level.

A1

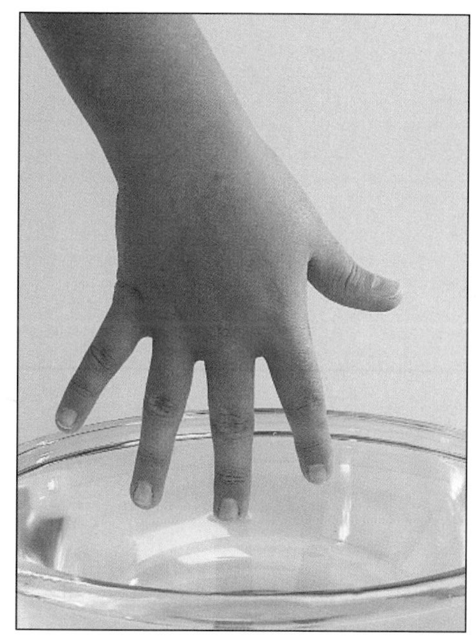

Nicky is lowering her hand into water.

List these cross-sections of her hand in the right order. C, A, B, D

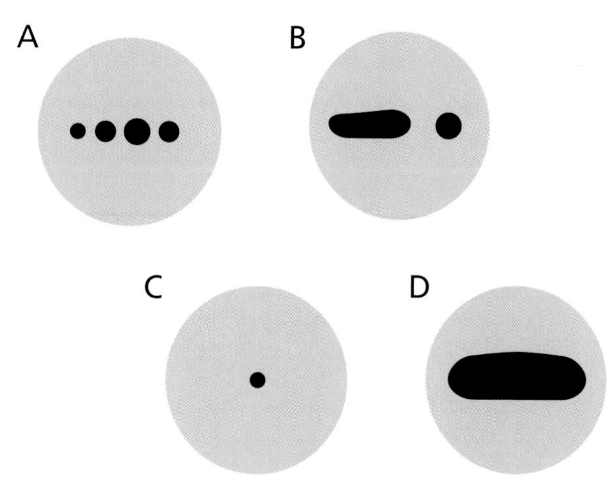

A2

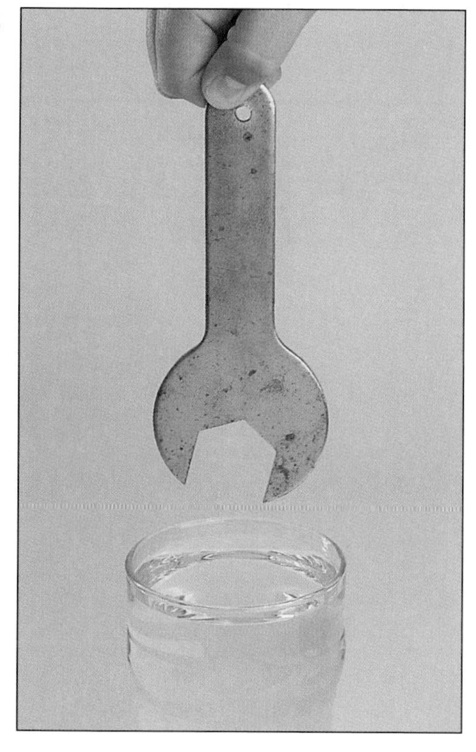

This spanner is being lowered into water.

List these cross-sections in the right order.

B, A, D, C

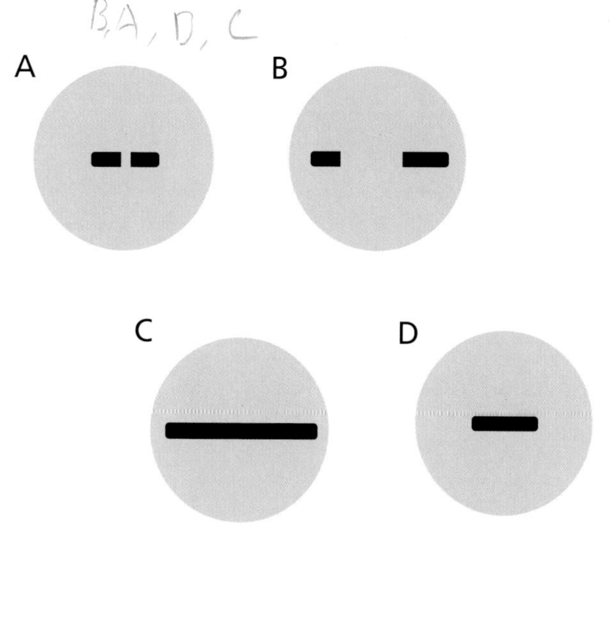

A3 Sketch a set of cross-sections for each of these.

Garden trowel

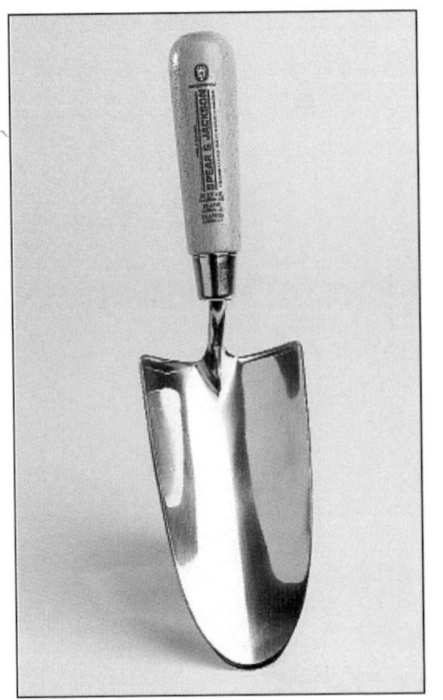

Garden hoe blade

Mallet

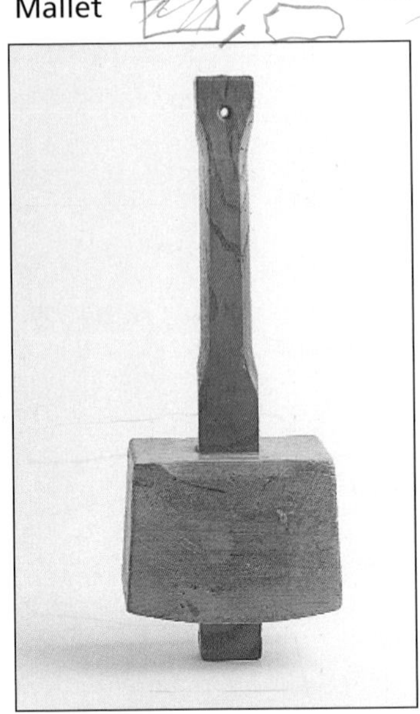

A4

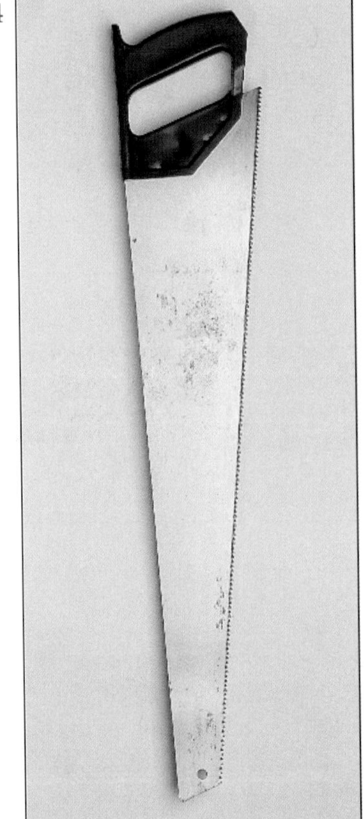

This saw is going to be lowered into water.

List these cross-sections in the right order.

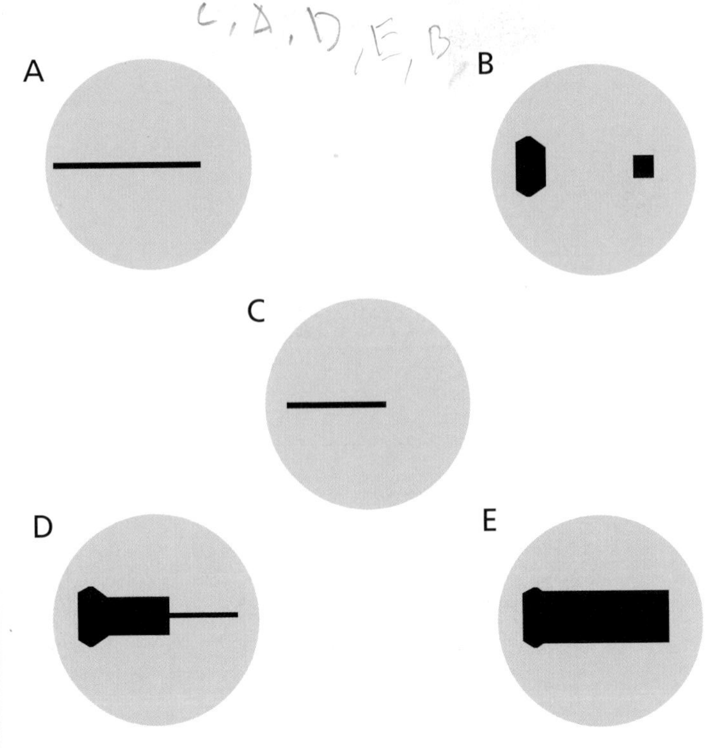

A5 This block is lowered into the water in three different ways.

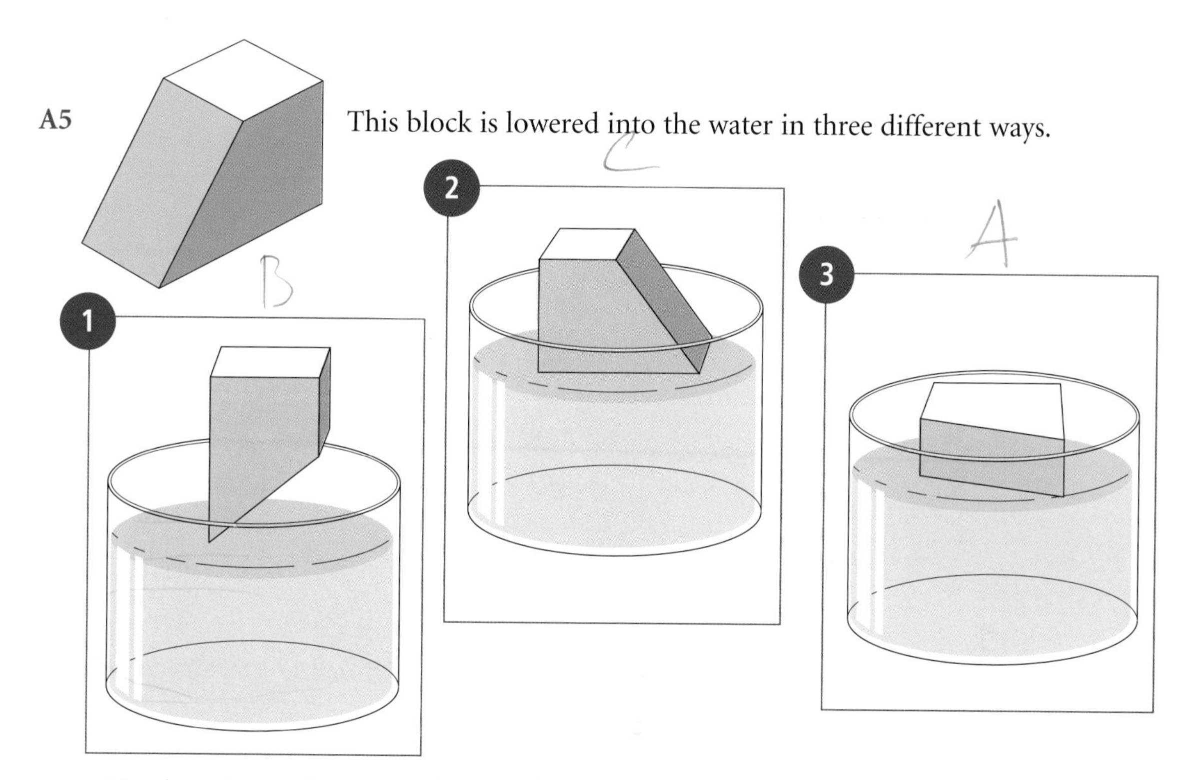

Match each set of cross-sections to the correct picture.

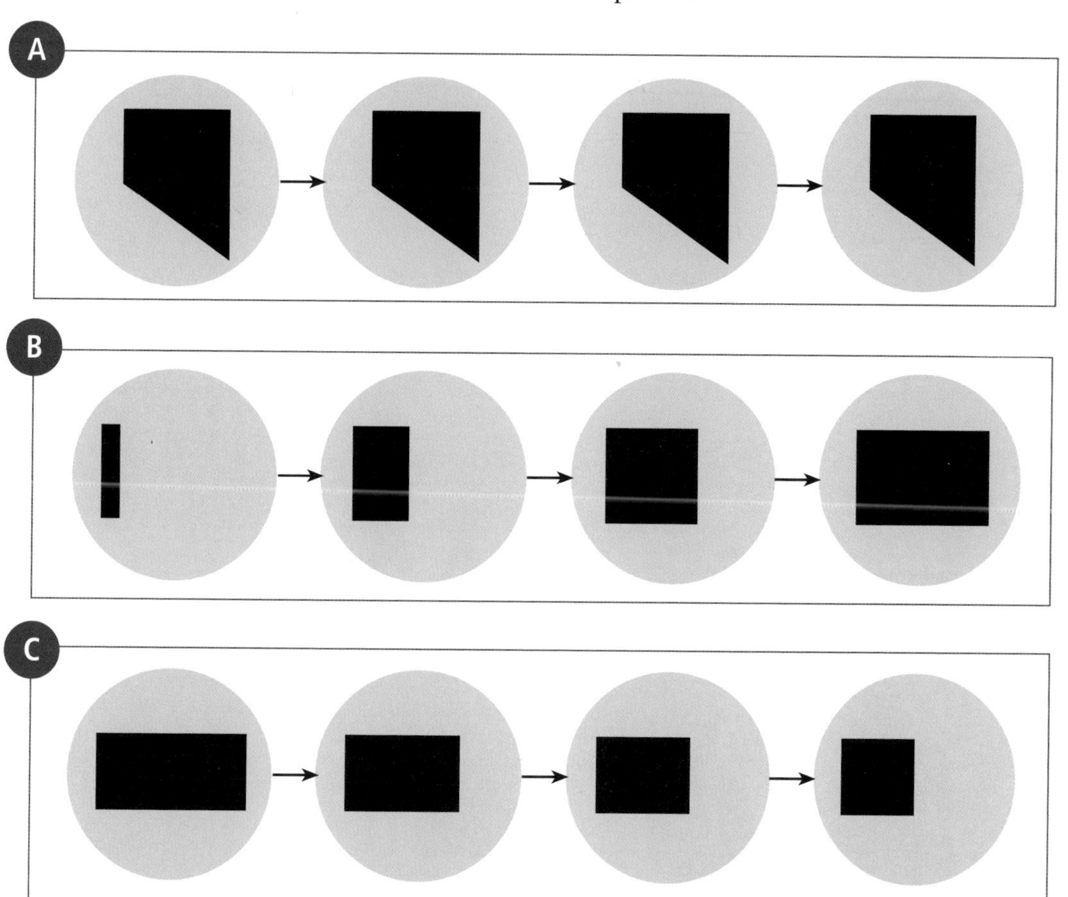

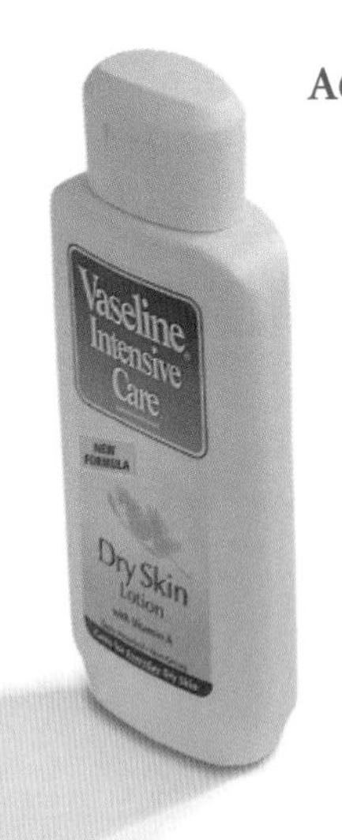

A6 This jar is being lowered into water in two different ways.

Sketch a set of cross-sections for each of the ways.

(a) 

(b)

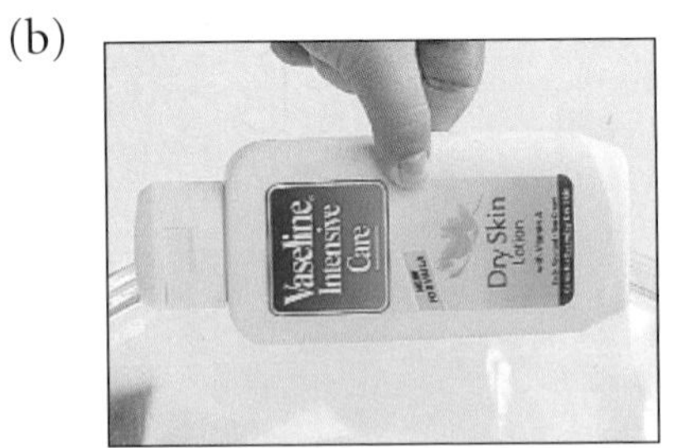

A7 Here are three different jars.

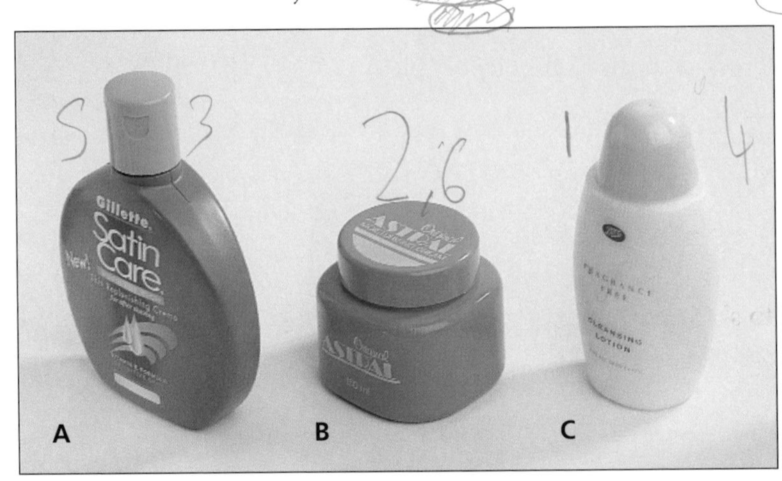

Which jar does each of these cross-sections belong to?

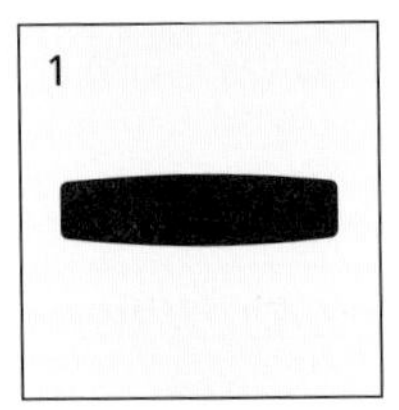

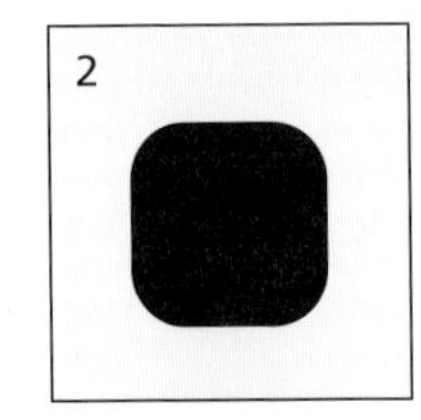

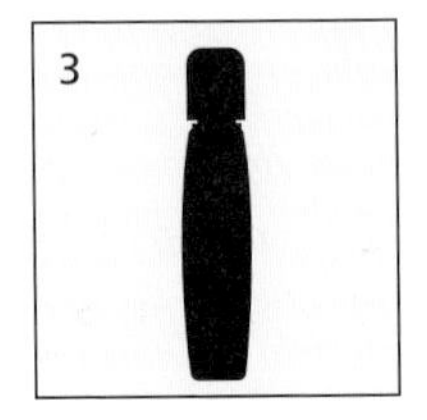

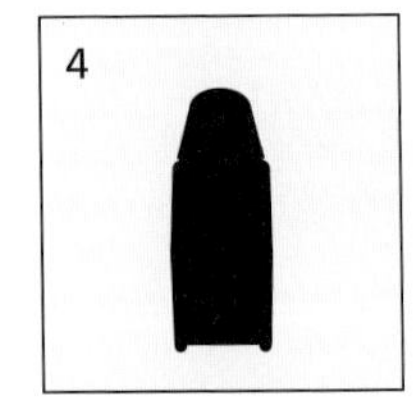

A8 (a) What object do you think each of these sets of cross-sections shows?

(i)

(ii)

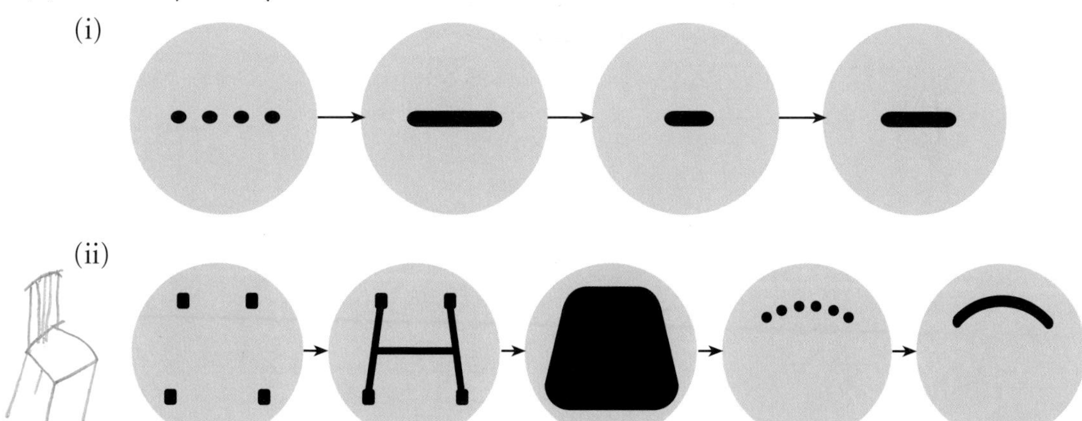

(b) Choose your own object and make up a similar puzzle.

***A9** The diagram shows two ways to cut through a cube to get different cross-sections.

(a) Draw sketches to show how to cut through the cube to get these cross-sections.

(i) 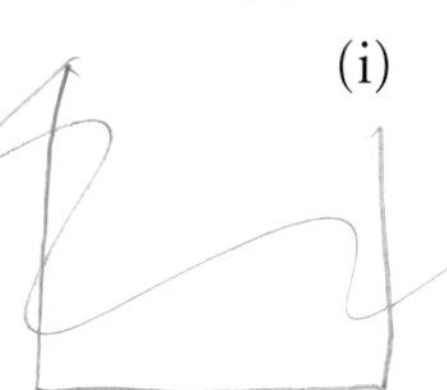

(ii) 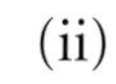

(iii) 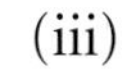

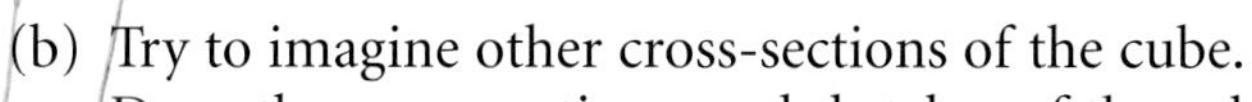

(b) Try to imagine other cross-sections of the cube.
Draw the cross-sections, and sketches of the cube showing how to get them.

B Planes of symmetry

The dotted line is a line of symmetry of this two-dimensional shape.

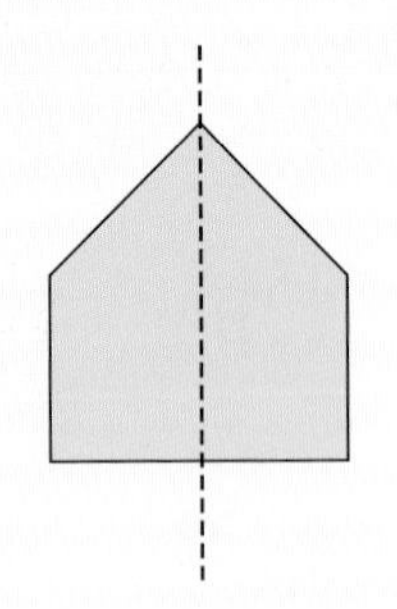

The shaded plane is a **plane of symmetry** of this three-dimensional shape.

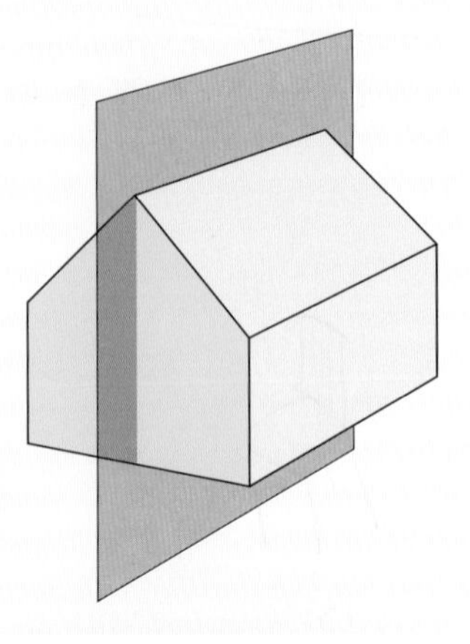

This shape has another plane of symmetry.

Where is it?

B1 The blue solid is a prism whose cross-section is an equilateral triangle.
One of its planes of symmetry is shown.

How many planes of symmetry does the prism have altogether?

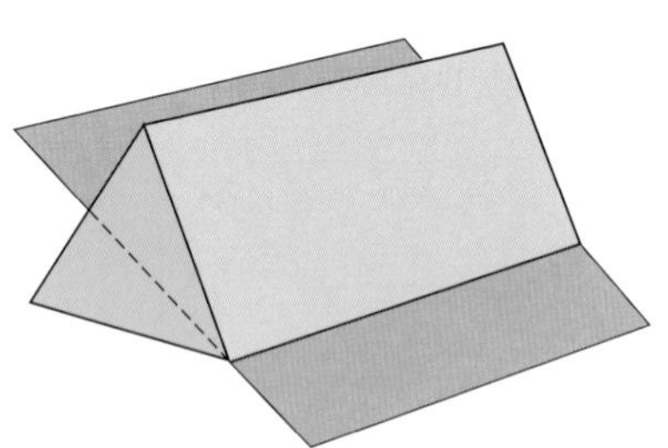

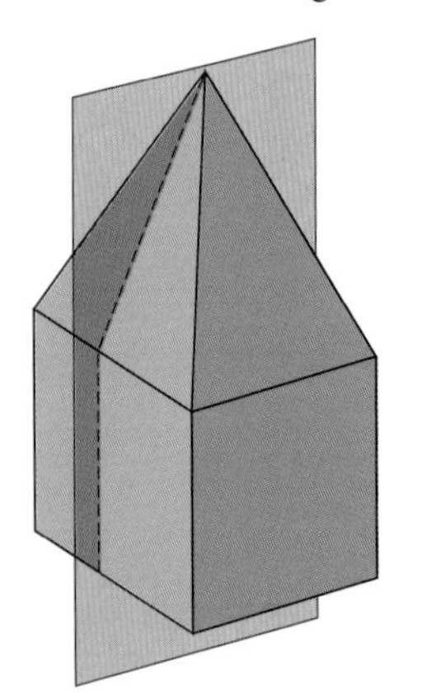

B2 The red solid consists of a square-based pyramid on top of a cube.
One of its planes of symmetry is shown.

How many planes of symmetry does the solid have altogether?

B3 How many planes of symmetry does this yellow cuboid have?

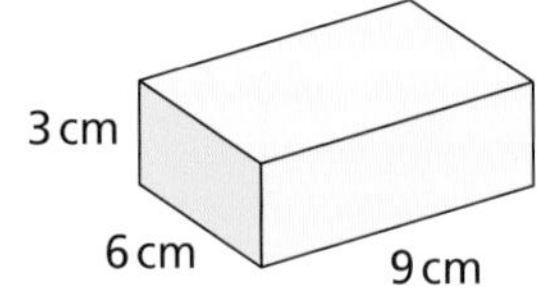

B4 Karen has two children's bricks, each like the yellow cuboid above.
She has arranged them to make this solid with one plane of symmetry.

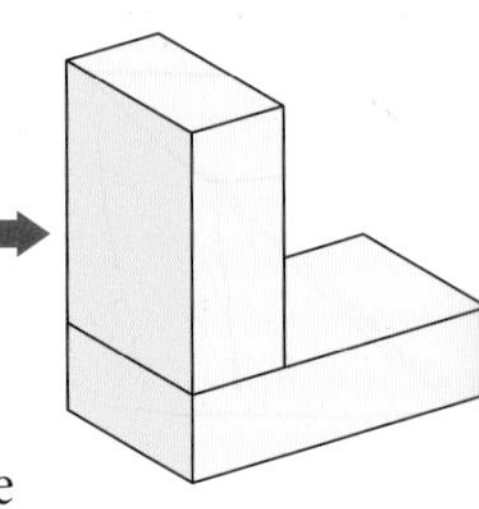

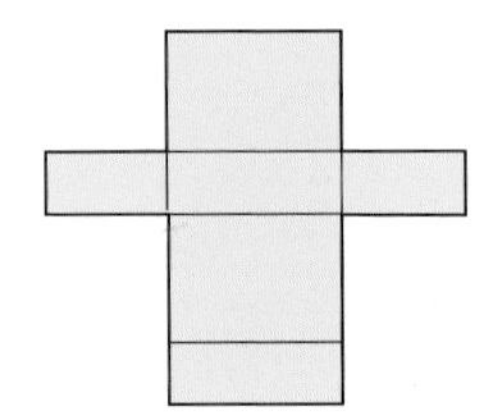

Draw the nets of the two cuboids on squared paper.
Cut them out and make the 'bricks'.

Draw sketches to show how to arrange the bricks to make a solid with

(a) 2 planes of symmetry (b) no plane of symmetry

(c) 3 planes of symmetry (d) 5 planes of symmetry

B5 How many planes of symmetry does a cube have?

What progress have you made?

Statement | **Evidence**

I can identify cross-sections of a solid.

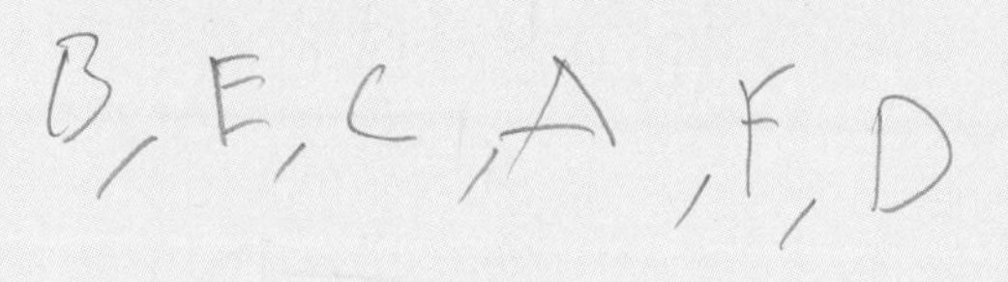

1 

This key is lowered into water.

List these cross-sections in the correct order.

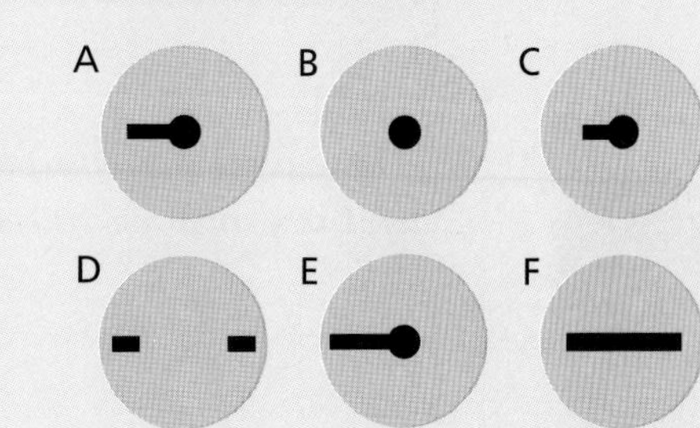

I can draw cross-sections of a solid.

2 Draw a set of four or five cross-sections of these as they are lowered into water.

(a) 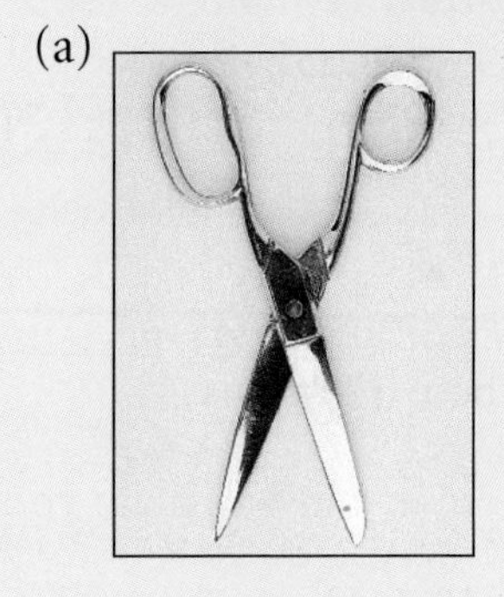

(b) 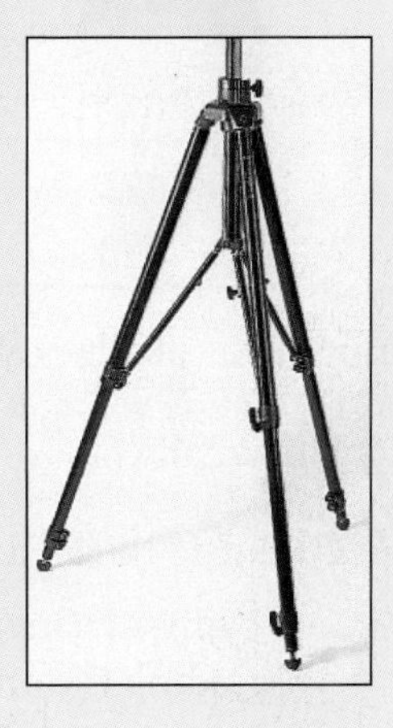

I can identify the planes of symmetry of a solid.

3 How many planes of symmetry does this square-based prism have?

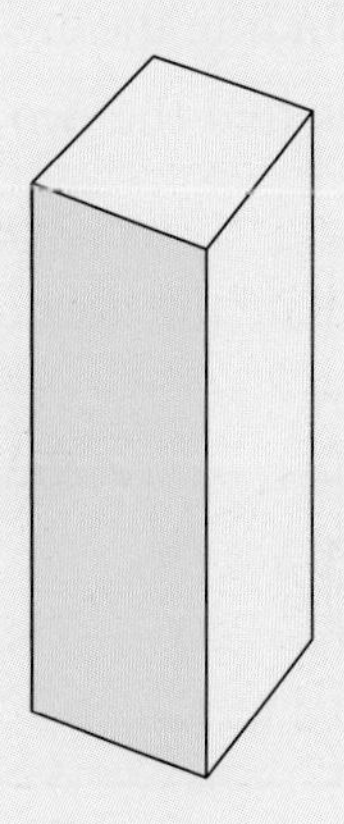

6 Units

This work will help you

- understand and use metric units of area, volume and capacity
- change from common imperial units to metric units

A Area

There are 100 cm in 1 m.
So there must be 100 cm² in 1 m².

That can't be right!
This area is 25 cm².
Four of these won't make 1 square metre.

Do not use a calculator for these questions.

A1 How many cm² are there in 1 m²? Explain your answer.

A2 Calculate the area of this rectangle in two different ways:

(a) by first changing both dimensions to cm and finding the area in cm²

(b) by first changing both dimensions to metres and finding the area in m²

Check that the two results are equivalent.

1.4 m
60 cm

A3 Calculate the area of each of these shapes. Give each result in both m² and cm².

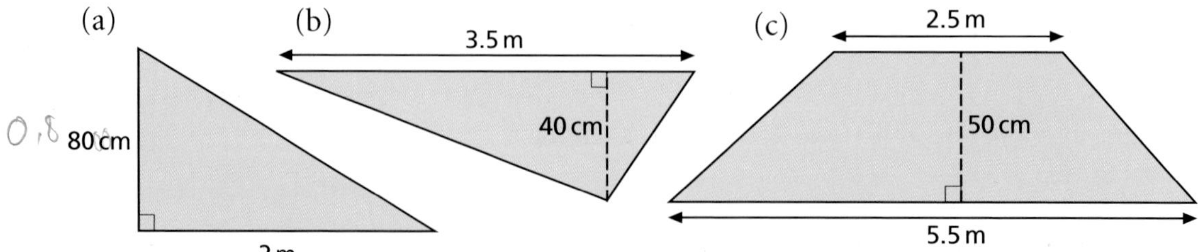

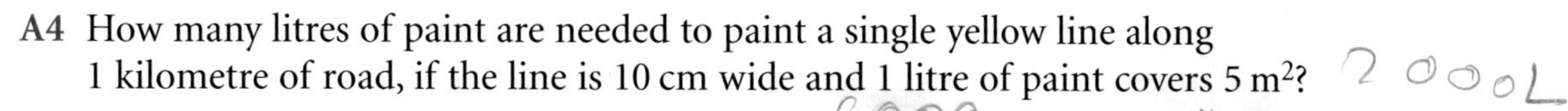

A4 How many litres of paint are needed to paint a single yellow line along 1 kilometre of road, if the line is 10 cm wide and 1 litre of paint covers 5 m^2?

A5 (a) Calculate the area of the sticky tape on each of these rolls.

(b) Which roll gives you more tape for your money? Explain.

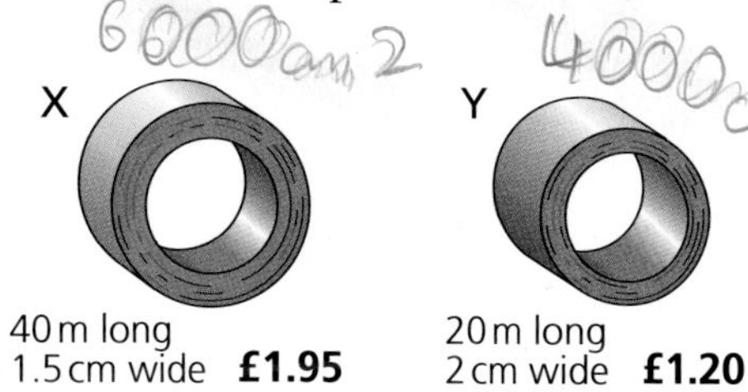

B Capacity and volume

Each of these containers holds 1 litre.

A hollow cube 10 cm by 10 cm by 10 cm holds 1 litre.

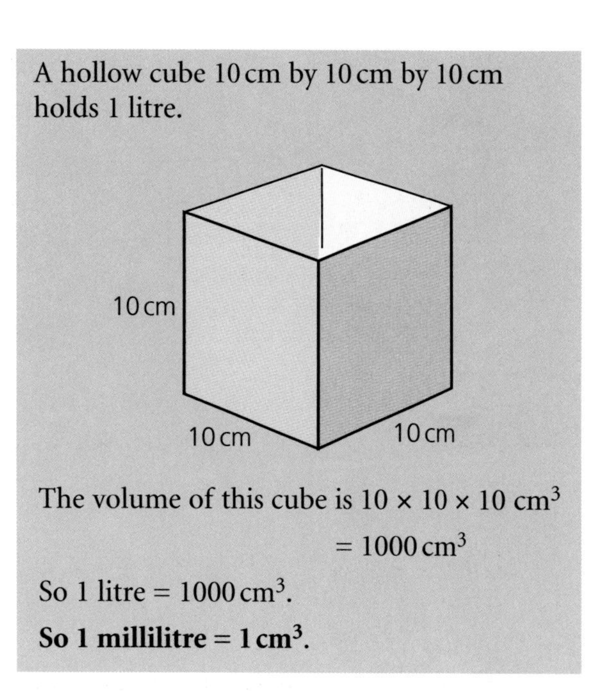

The volume of this cube is $10 \times 10 \times 10\ cm^3$

$= 1000\ cm^3$

So 1 litre = 1000 cm^3.

So 1 millilitre = 1 cm^3.

B1 How many litres of liquid would it take to fill each of these?

(a) Eight 250 ml cartons of orange juice

(b) Twelve 750 ml bottles of wine

(c) A carton 20 cm by 20 cm by 50 cm

(d) A petrol tank 70 cm by 80 cm by 120 cm

B2 The capacity of a car engine is measured as either litres or cc (cubic centimetres, cm^3). What is the size in litres of each of these engines?

(a) 2000 cc (b) 3500 cc (c) 850 cc

B3 Write these quantities as decimals of a litre.

(a) 250 ml (b) 1 cm^3 (c) 50 ml

(d) 75 cl (1 cl = 1 centilitre = 10 ml)

B4 Abby says, 'There are 100 cm in 1 m, so there must be 100 cm^3 in 1 m^3'.

Derek says, 'That can't be right, because this cuboid's volume is 100 cm^3 and it's nowhere near the size of a cubic metre.'

How many cm^3 are there in 1 m^3?
Explain your answer.

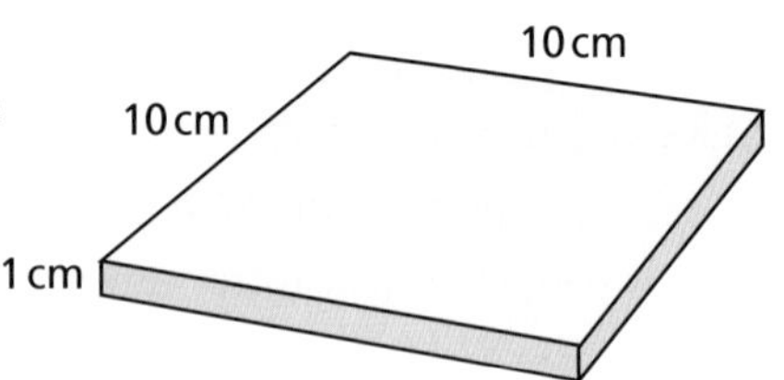

B5 An aquarist has three types of additive for her fish tanks.

Anti Fungus and Fishrot	add 1 teaspoon per 50 litres
Interpret Aqua Salt	add 1 teaspoon per 25 litres
Tetra Aqua Safe	add 5 ml for every 10 litres of water

How much of each additive would she need to add to these tanks?

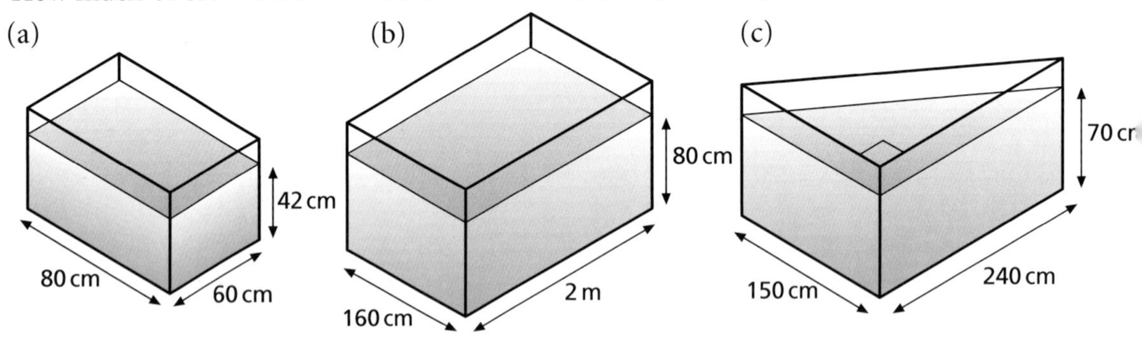

B6 (a) How many litres are there in 1 m^3?

(b) A fire engine has a tank which is a cuboid 1.5 m long by 0.8 m wide and 1 m tall. How deep must the water be for the tank to contain 1000 litres?

B7 A swimming pool measures 15 m by 10 m by 0.8 m.

(a) How many litres of water does it take to fill it?

(b) A normal tap flows at a rate of about 12 litres per minute.
Roughly how long would it take to fill a swimming pool of this size?

(c) Chlorine is added daily to the water at a concentration of 1 g per 40 litres.
How much chlorine needs to be added to this pool?

B8 A 2 litre carton is made from a cuboid with a square base.
If the carton is 25 cm high, what is the width of the base?

Design a carton

Sketch a net that would make a carton with capacity 500 ml.

What area of card is needed to make your carton?

What is the ideal shape to make a 500 ml carton to use the least amount of card? Investigate.

C Imperial units

The old system of 'imperial' units is still used sometimes.
These rough rules for converting to metric units are good enough for most purposes.

To change from miles to km, divide by 5 and multiply by 8.
So 30 miles is roughly $\frac{30}{5} \times 8 = 48$ *km.*

A foot is about 30 cm.

A kilogram is about 2.2 pounds (lb).
So 1 lb is less than $\frac{1}{2}$ *kg.*

A litre is about $1\frac{3}{4}$ *pints.*
A pint is about $\frac{3}{5}$ *litre.*

A gallon is about $4\frac{1}{2}$ *litres.*

Do these questions without a calculator.

C1 Change the distances given in km on these road signs to miles.

Calais	16
Boulogne	24

Firenze	64
Siena	32

Limerick	52
Ennis	20

C2 Convert these speed limits in miles per hour to kilometres per hour.

30

C3 A company which supplies oil for central heating has been selling in gallons.
It needs to convert its bills to litres.

How many litres are there in each of these?

(a) 8 gallons (b) 20 gallons (c) 50 gallons (d) 200 gallons

C4 Regular blood donors keep logbooks which show how many pints of blood they have given. Change these into litres.

(a) 21 pints (b) 45 pints (c) 120 pints

C5 A yard is equal to 3 feet.
A cricket pitch is 22 yards long. How long is this in metres, roughly?

C6 (a) Rewrite the rule for changing miles to kilometres as a rule for changing kilometres to miles.

(b) Change these to miles.

(i) 48 km (ii) 200 km (iii) 60 km (iv) 144 km

C7 An old book says that the main room of a ruined house had a square floor 10 feet by 10 feet.

(a) Change 10 feet to metres, roughly.

(b) What is the area of the floor, roughly, in m^2?

(c) In **square feet**, the area of the floor is $10 \times 10 = 100$ sq ft.
Roughly how many square feet are there in a square metre?

D Mixed questions

D1 How many 250 millimetre pieces can be cut from a 3 metre plastic strip?

D2 A box which weighs 310 g contains 24 packs of cheese each weighing 125 g.
How much do the box and the pack weigh altogether, in kilograms?

D3 How many 50 milligram tablets can be made from 300 grams of vitamin C?

D4 Petra has to take a teaspoonful of medicine four times a day.
A teaspoon holds 5 millilitres.
For how many days should a 0.5 litre bottle of medicine last?

D5 A wine glass holds 125 ml.
How many glasses can be filled from a 0.75 litre bottle of wine?

D6 1 litre of water weighs 1 kg.

This fish tank weighs 9 kg when empty.
How much does it weigh when filled with water to a depth of 30 cm?

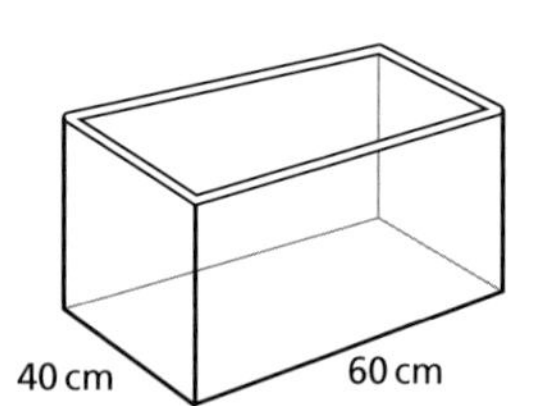

D7 An adult's body contains about 8 pints of blood.
How much is this in litres, roughly?

D8 Americans measure their weight in pounds.
What is the weight in kilograms, roughly, of a man weighing 180 pounds?

D9 An inch is about $2\frac{1}{2}$ cm.
A skirt is described as 'length 24 inches, waist 21 inches'.
What are these measurements in centimetres, roughly?

What progress have you made?

Statement	Evidence
I know the connection between litres, millilitres and cm^3.	**1** How many litres are there in (a) 2300 cm^3 (b) 250 ml (c) a carton 25 cm by 15 cm by 30 cm
I can solve problems using capacity.	**2** A bath is roughly a cuboid 55 cm wide, 35 cm deep and 1.5 m long. How long would it take to fill at 12 litres per minute?
I know the rough metric equivalents of common imperial units.	**3** Roughly how many (a) kilometres are there in 100 miles (b) kilograms are in a stone (14 pounds)

7 Simplifying expressions

This work will help you
- simplify algebraic expressions such as $5x - 3 - 2x + 1$
- use algebra to solve problems

A Simplifying

Set 1

A $1 + x - 2x + 3$

B $5 - x$

C $2x + 5 - x$

D $6 + 2x - 1 - 3x$

Set 2

A $1 - 5y - 2y + 5$

B $6 - 7y$

C $6 - 3y$

D $2 - 5y + 2y + 4$

Set 3

A $2x + y - x + 5y$ B $x - 4y + 2x - 2y$ C $5 + 3x - 7x + 5x + 2$

D $7 + 3y + x - 3y$ E $5x + y - 2x - 7y$ F $8y + x - 2y$

A1 Simplify these expressions.

(a) $6p + 3p - 5p$ (b) $7q - 3q + q$ (c) $2r + 6 - r + 7$

(d) $4s + 5 + 3s - 2$ (e) $3t + 5 + t - 8 + 2t$ (f) $8u - 4u - 2u$

(g) $5v - 4 - 2v + 3$ (h) $6 - 2w + 5 + 3w$ (i) $6x - 8x + 5x$

A2 Find four matching pairs of equivalent expressions.

A $11 - 3k - 3 - k - 1 - 6k$ B $1 + 2k - 5 - 12k + 13$

C $7 + 4k$ D $5 + 6k + 2 - 5k - 3k$ E $5 - 10k + 4$

F $7 - k - k$ G $7 - 10k$ H $4 - 3k + 3 + 7k$

A3 Simplify these expressions.

(a) $5 + 7k - 9k$ (b) $3 + 4j - 6 + j$ (c) $6 + 2h - 3h - 5$

(d) $10 + 4g - 5g + 3g$ (e) $9 - 4f - 2 + 3f$ (f) $1 - e - 2e + 5$

(g) $8d + 8 - 2d - 12$ (h) $3c - 8 - 6c + 10$ (i) $7 + 2b - 4b + 8b - 9b$

A4 Simplify these expressions.

(a) $3z + 5y + z + 3y$ (b) $2x + 5w + x - 3w$ (c) $5 + 5u + 2v - u + 2$

(d) $6s + 2t - 7s + t$ (e) $6 + 3q - 5r + 2q - 4$ (f) $2n + 5p - 3n - 2p + 6n$

(g) $4l + 6 - 3m - 3l$ (h) $10 - 3k - j - 5k - 2j$ (i) $2g + 5h - 6g - h$

(j) $e + f - 3 + 2e - 5f$ (k) $2 + c - 5 + 2c - d$ (l) $3a - 6b + 5a - b + 3b$

B Magic squares

In a 'magic square' the numbers in each **row**, **column** and **diagonal** add up to the same total (the 'magic total').

The magic total for the square on the right is 24.

11	6	7
4	8	12
9	10	5

Totals
11 + 6 + 7 = 24
4 + 8 + 12 = 24
9 + 10 + 5 = 24
11 + 8 + 5 = 24

B1 Copy and complete these magic squares.

(a)

31		
	21	30
		11

(b)

8			14
21		3	
	13		11
12	5	17	

B2

6	1	2
⁻1	3	7
4	8	0

This is not a magic square.

How could you turn it into a magic square by changing one number?

B3 (a) Make a grid of numbers by replacing each expression in grid V with its value when $v = 2$.

(b) Make a grid of numbers by replacing each expression with its value when $v = 3$.

(c) Which of your grids is a magic square?

Grid V

5 + 2*v*	8 − *v*	*v* + 4
3*v* − 2	5 + *v*	5*v*
10 − *v*	*v* + 6	2*v* + 1

Grid W

w + 2	3*w*+8	5 − *w*
8 − *w*	5 + *w*	3*w*+2
5+3*w*	2 − *w*	*w* + 8

B4 (a) Simplify this expression for the total of the first row of grid W: $w + 2 + 3w + 8 + 5 - w$

(b) Find the total of the three expressions in each of the other rows, each column and each diagonal.

(c) Explain why the grid will be a magic square for **any** value of w.

(d) What is the magic total when $w = 1$?

B5 Use grid W to find

(a) a magic square with 8 in the centre (b) a magic square whose magic total is 30

B6 (a) Copy and complete grid X to make a magic square.

(b) Use your grid to make a magic square with a magic total of 18.

Grid X

$2x + 1$	$2 + x$	$6x - 3$
		$4 - x$

Grid Y

	$3 - 4y$	
	$5 - 2y$	
	7	$2 - y$

B7 (a) Copy and complete grid Y to make a magic square.

(b) When $y = 1$, what is the magic total for the square?

B8 (a) Copy and complete grid Z to make a magic square.

(b) Make a magic square by replacing each expression by its value when $y = 5$ and $z = 2$.

Grid Z

		$6y$
	$3y + z$	$5z - y$
		$4y - 2z$

Grid A

$5a + 11b - 1$	$16b + 4a - 7$	$7 - 20b$	$1 + 3a - 7b$
$7 + 4a$			$3a - 1$
		$4a$	
	$5a$	$4b + 3a$	

***B9** (a) Copy and complete grid A to make a magic square.

(b) Use grid A to make a magic square with a magic total of 24.

Grid B

$c - 3d$	$22c - 12d$	$6d + 2c$	$9c + 3d$
$9d - 5c$	$16c$	$8c - 6d$	$15c - 9d$
$20c - 8d$	$d - c$	$11c - d$	$4c + 2d$
$18c - 4d$	$5d - 3c$	$12c - d$	$6c - 2d$

***B10** (a) Show that grid B is not a magic square.

(b) How could you change one expression to make it a magic square?

(c) What is the magic total when $c = 1$ and $d = 4$?

C Walls

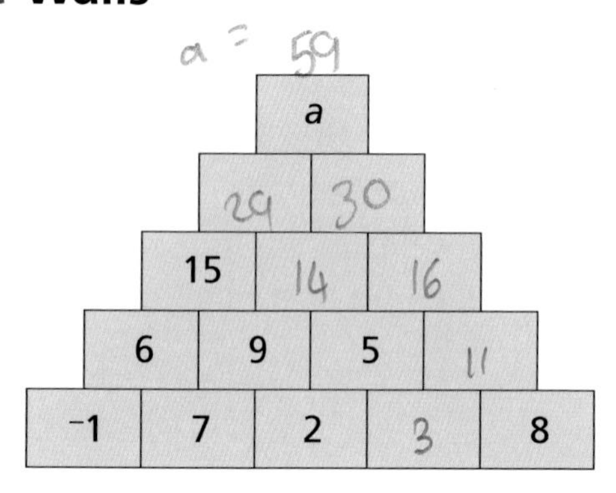

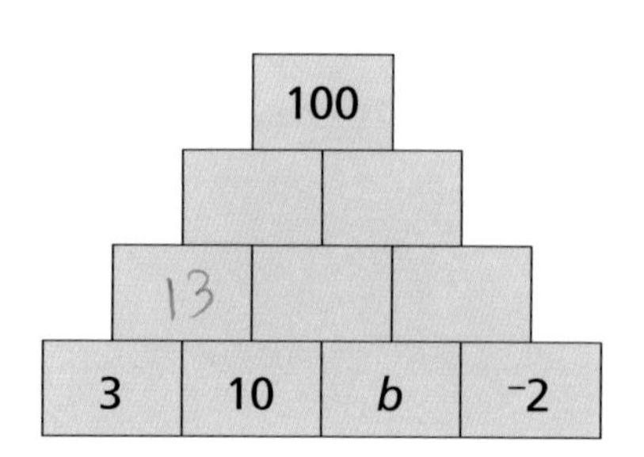

C1 What is the value of the letter in each of these walls?

(a)

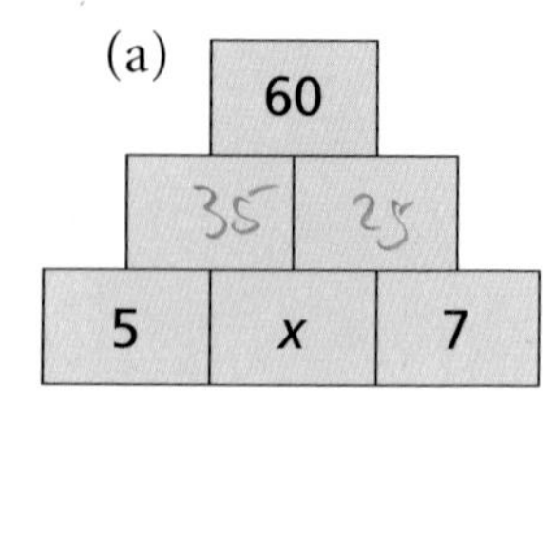

(b)

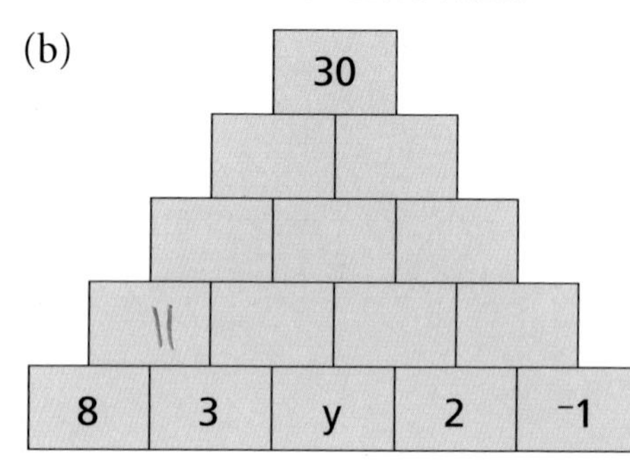

(c)

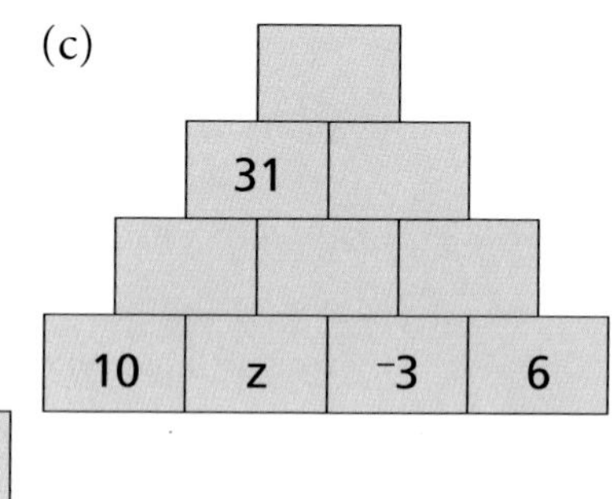

C2 Find all the missing numbers for each wall.

(a)

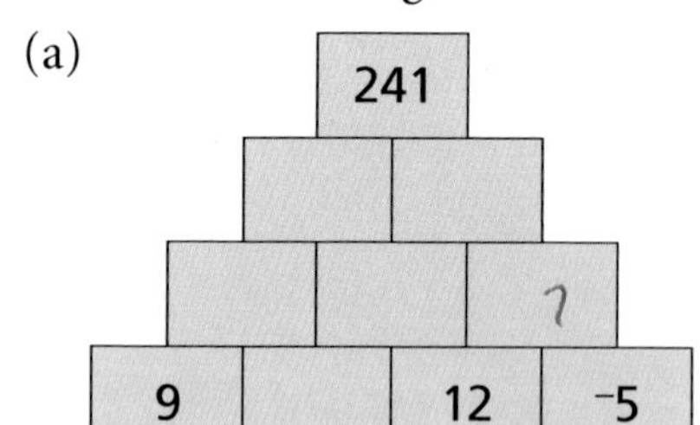

(b)

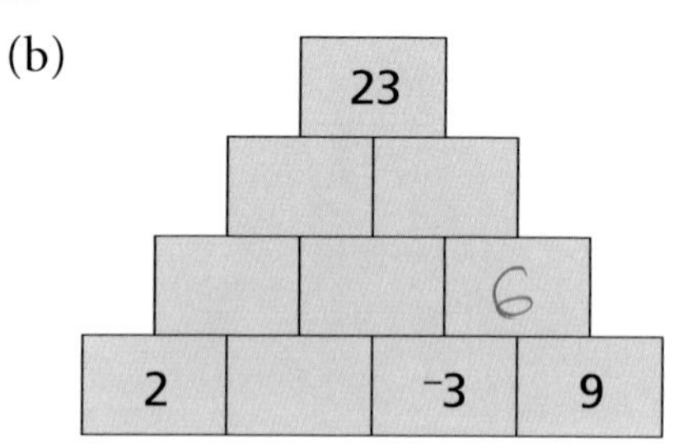

(c)

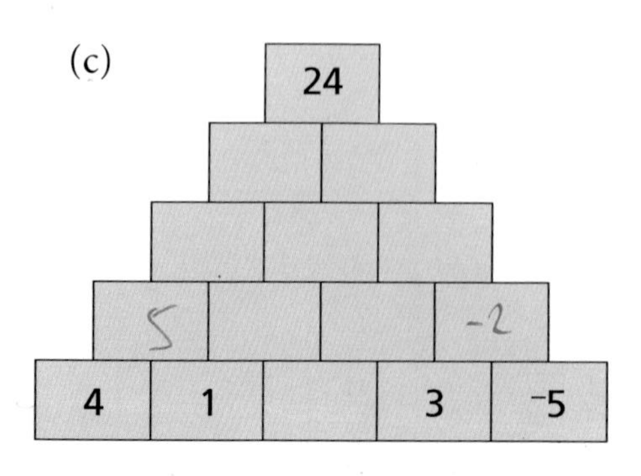

(d)

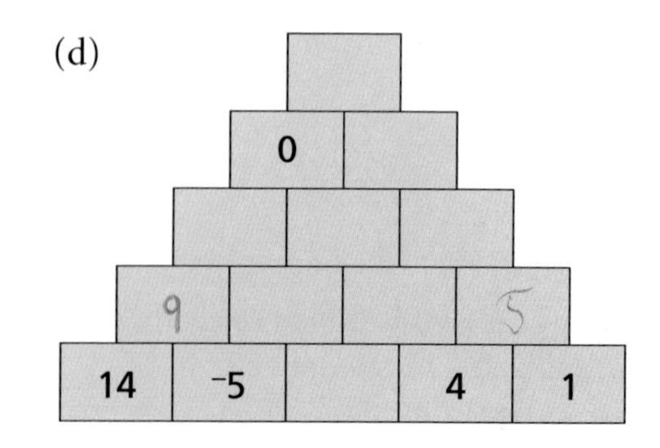

C3 In this wall, all the numbers on the bottom row are the same.

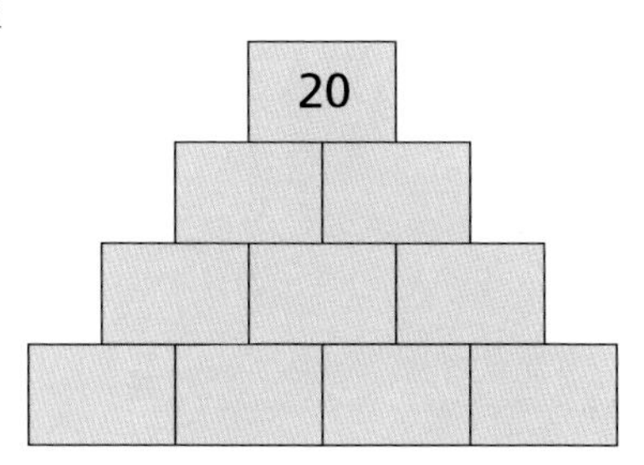

What are the numbers on the bottom row?

*C4 (a) Find an expression for each brick in this wall.

(b) What value for h gives 100 in the top brick?

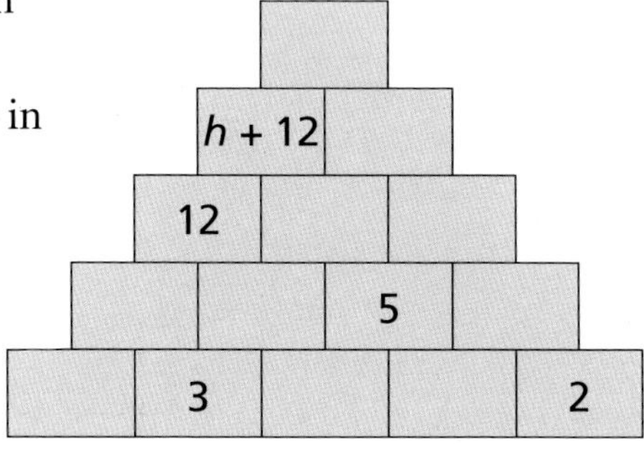

*C5 Find a value for n so that the numbers in the green bricks are the same.

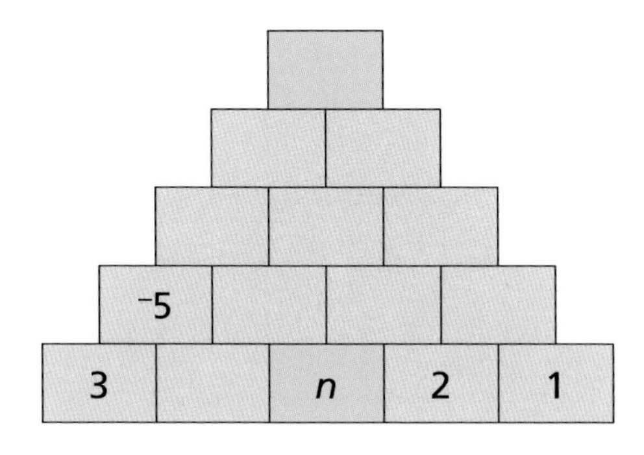

*C6 Find a value for y so that the numbers on the top bricks of these walls are equal.

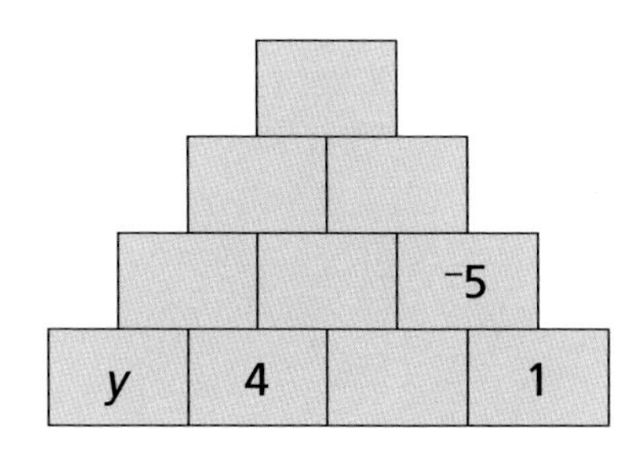

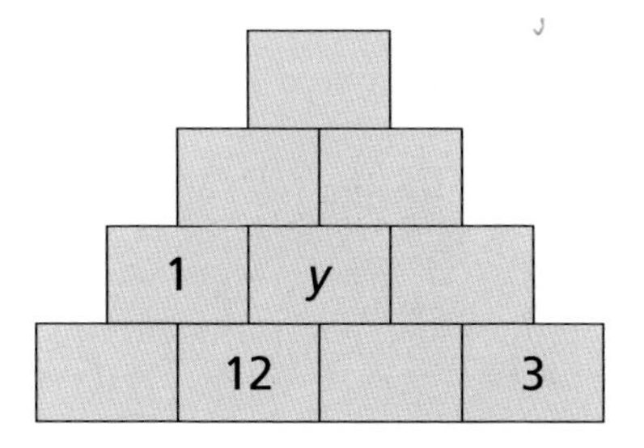

What progress have you made?

Statement	Evidence

Statement

I can simplify expressions like $2x + 3y - x + y + 4$.

Evidence

1 Simplify the following expressions.

(a) $2p + 5p - 3p$

(b) $4 - q + 1 + 7q$

(c) $6 + 3r - 5r - 1$

(d) $8 - 3s + 1 + 7s - 10$

(e) $3w + 5x - w - 7x$

(f) $4 + y - 5 + 3y - 2z$

Statement

I can use algebra to solve problems.

Evidence

2 (a) Copy and complete these grids to make magic squares.

(i)

$5x + 2$	3	$7x - 2$
$6x - 3$		

(ii)

$7a - 2b$	$2a - 3b$	
	$4a - b$	
	$6a + b$	

(b) Choose one of these grids to make a magic square with 9 in the centre square.

Show all your working.

3 Find all the missing numbers for this wall.

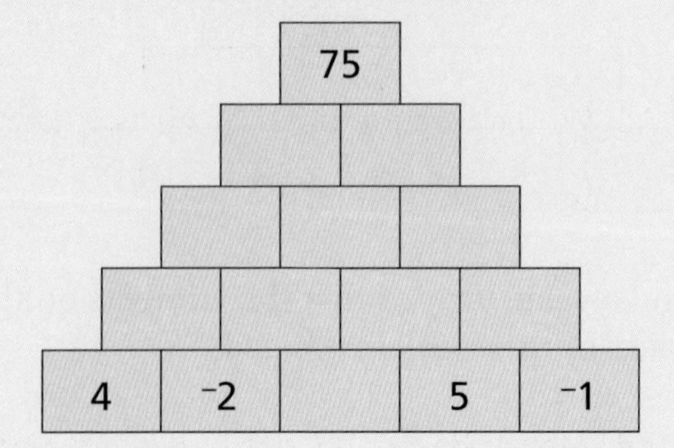

8 Fractions and decimals

This work will help you

- revise ordering, adding and subtracting fractions
- multiply and divide a whole number by a fraction
- convert between fractions and decimals, including recurring decimals

A Simplifying and comparing fractions

A1 Copy and complete these pairs of equivalent fractions.

(a) $\frac{4}{5} = \frac{?}{20}$ (b) $\frac{12}{40} = \frac{?}{10}$ (c) $\frac{3}{8} = \frac{?}{24}$ (d) $\frac{2}{5} = \frac{10}{?}$ (e) $\frac{16}{30} = \frac{?}{90}$

(f) $\frac{2}{7} = \frac{10}{?}$ (g) $\frac{8}{9} = \frac{32}{?}$ (h) $\frac{4}{11} = \frac{32}{?}$ (i) $\frac{12}{20} = \frac{?}{5}$ (j) $\frac{1}{5} = \frac{?}{25}$

A2 Simplify these fractions as far as possible.

(a) $\frac{18}{20}$ (b) $\frac{9}{21}$ (c) $\frac{25}{45}$ (d) $\frac{18}{48}$ (e) $\frac{24}{80}$

(f) $\frac{24}{60}$ (g) $\frac{8}{32}$ (h) $\frac{20}{90}$ (i) $\frac{36}{84}$ (j) $\frac{30}{56}$

To compare fractions, change them to the same denominator.

Example Which is greater, $\frac{3}{5}$ or $\frac{5}{8}$?

$\frac{3}{5} = \frac{24}{40}$ $\frac{5}{8} = \frac{25}{40}$ So $\frac{5}{8}$ is greater.

A3 Which fraction in each pair is greater?

(a) $\frac{3}{4}$ or $\frac{11}{12}$ (b) $\frac{3}{4}$ or $\frac{7}{9}$ (c) $\frac{5}{7}$ or $\frac{7}{10}$ (d) $\frac{3}{8}$ or $\frac{5}{12}$ (e) $\frac{9}{11}$ or $\frac{5}{6}$

(f) $\frac{5}{8}$ or $\frac{2}{3}$ (g) $\frac{3}{10}$ or $\frac{2}{7}$ (h) $\frac{4}{5}$ or $\frac{7}{9}$ (i) $\frac{3}{5}$ or $\frac{2}{3}$ (j) $\frac{11}{16}$ or $\frac{2}{3}$

A4 Albert, Bess and Charlie share two identical pizzas between them.
Albert has $\frac{3}{5}$ of the first pizza and gives the rest to Charlie.
Bess has $\frac{2}{3}$ of the second pizza and gives the rest to Charlie.

(a) Who has most pizza? Explain how you get your answer.

(b) Who has least? Explain your answer.

B Adding and subtracting fractions

B1 Work these out. Simplify the result where possible.

(a) $\frac{3}{4} - \frac{2}{3}$ (b) $\frac{1}{12} + \frac{2}{3}$ (c) $\frac{3}{5} - \frac{1}{5}$ (d) $\frac{3}{8} + \frac{1}{5}$ (e) $\frac{5}{6} - \frac{1}{4}$

(f) $\frac{5}{6} - \frac{3}{5}$ (g) $\frac{1}{8} + \frac{1}{5}$ (h) $\frac{3}{5} + \frac{1}{4}$ (i) $\frac{7}{8} - \frac{2}{3}$ (j) $\frac{5}{12} - \frac{1}{4}$

B2 Work these out.

(a) $\frac{1}{4} + \frac{1}{5} + \frac{1}{6}$ (b) $\frac{1}{3} + \frac{1}{5} + \frac{1}{8}$ (c) $\frac{1}{3} + \frac{1}{5} - \frac{1}{6}$ (d) $\frac{1}{4} - \frac{1}{5} + \frac{1}{6}$

B3 Which two of these fractions do you add together to get the largest possible result? Explain your answer.

$\frac{3}{4}$ $\frac{3}{5}$ $\frac{5}{8}$ $\frac{7}{10}$

B4 What fraction is missing in each of these?

(a) $\frac{1}{5} + \textbf{?} = \frac{3}{4}$ (b) $\textbf{?} + \frac{1}{6} = \frac{3}{8}$ (c) $\textbf{?} - \frac{1}{5} = \frac{1}{6}$ (d) $\frac{1}{4} - \textbf{?} = \frac{1}{6}$

B5 (a) Work out $\frac{3}{5} + \frac{1}{4}$. (b) Hence work out $2\frac{3}{5} + 1\frac{1}{4}$.

B6 (a) Work out $\frac{3}{4} - \frac{1}{3}$. (b) Hence work out $4\frac{3}{4} - 2\frac{1}{3}$.

B7 Work these out. Simplify the result where possible.

(a) $3\frac{2}{3} + 1\frac{1}{6}$ (b) $3\frac{2}{3} - 1\frac{1}{6}$ (c) $4\frac{3}{4} + 1\frac{1}{8}$ (d) $1\frac{1}{3} - \frac{3}{4}$ (e) $1\frac{1}{4} - \frac{4}{5}$

C Multiplying a whole number by a fraction

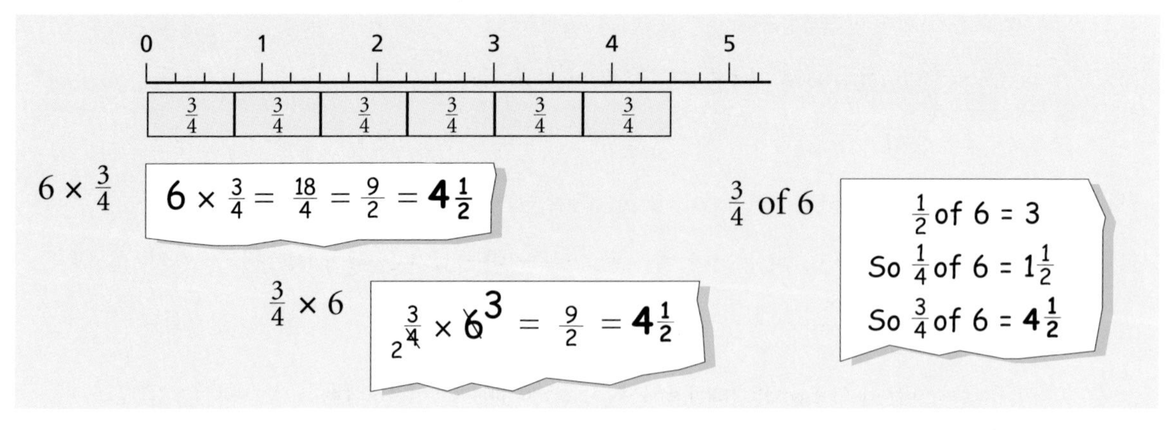

C1 Work these out.

(a) $10 \times \frac{1}{2}$ (b) $\frac{1}{2} \times 15$ (c) $12 \times \frac{1}{4}$ (d) $\frac{1}{4} \times 9$ (e) $2 \times \frac{3}{8}$

(f) $5 \times \frac{2}{3}$ (g) $\frac{3}{4} \times 14$ (h) $8 \times \frac{2}{5}$ (i) $\frac{5}{6} \times 15$ (j) $12 \times \frac{5}{8}$

C2 Work these out.

(a) $\frac{1}{3} \times 8$ (b) $9 \times \frac{2}{3}$ (c) $\frac{3}{8} \times 20$ (d) $16 \times \frac{5}{8}$ (e) $7 \times \frac{1}{3}$

(f) $10 \times \frac{2}{5}$ (g) $3 \times \frac{3}{4}$ (h) $\frac{7}{10} \times 5$ (i) $15 \times \frac{2}{5}$ (j) $\frac{5}{6} \times 7$

D Dividing a whole number by a fraction

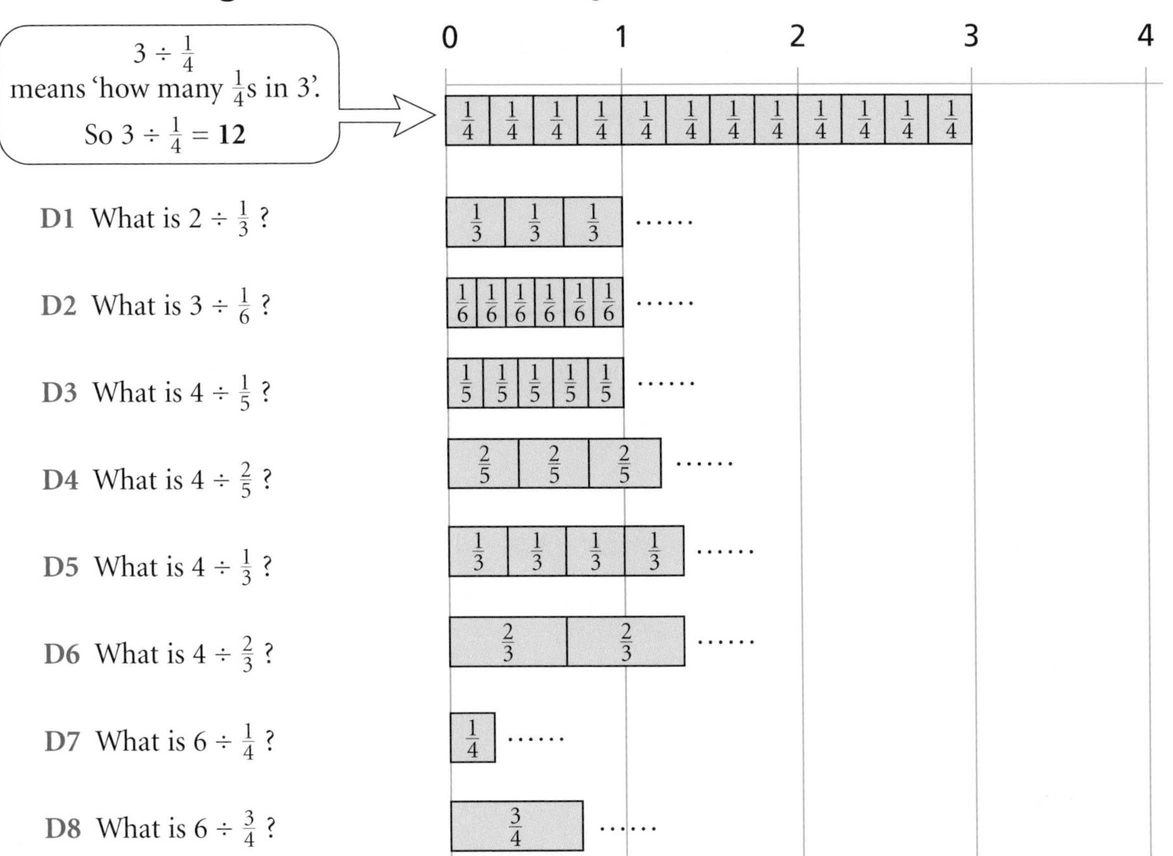

0 1 2 3 4
$3 \div \frac{1}{4}$ means 'how many $\frac{1}{4}$s in 3'.
So $3 \div \frac{1}{4} = \mathbf{12}$
D1 What is $2 \div \frac{1}{3}$?
D2 What is $3 \div \frac{1}{6}$?
D3 What is $4 \div \frac{1}{5}$?
D4 What is $4 \div \frac{2}{5}$?
D5 What is $4 \div \frac{1}{3}$?
D6 What is $4 \div \frac{2}{3}$?
D7 What is $6 \div \frac{1}{4}$?
D8 What is $6 \div \frac{3}{4}$?

You can do these in a similar way, by imagining the diagrams.

D9 (a) $9 \div \frac{1}{4}$ (b) $9 \div \frac{3}{4}$ (c) $6 \div \frac{1}{3}$ (d) $6 \div \frac{2}{3}$ (e) $10 \div \frac{1}{8}$ (f) $10 \div \frac{5}{8}$

D10 (a) $6 \div \frac{1}{5}$ (b) $6 \div \frac{2}{5}$ (c) $6 \div \frac{3}{5}$

D11 (a) Perry writes down a rule for dividing a whole number by a fraction.
It starts like this.

> Example: $4 \div \frac{2}{3}$
>
> First multiply the whole number by the denominator of the fraction: 4 x 3 = 12.
> Then

Finish the rule.

(b) What do you get if you use the rule for $2 \div \frac{3}{4}$?
Draw a diagram to show that the result is correct.

D12 Work these out.

(a) $8 \div \frac{1}{4}$ (b) $\frac{1}{4} \times 8$ (c) $12 \times \frac{2}{3}$ (d) $12 \div \frac{2}{3}$ (e) $7 \times \frac{1}{8}$ (f) $7 \div \frac{1}{8}$

E Changing fractions to decimals

To change a fraction to a decimal, you divide the numerator by the denominator.

For example, $\frac{5}{8} = 5 \div 8$
$= \mathbf{0.625}$

$$\begin{array}{r} 0.6\ 2\ 5 \\ 8\overline{)5.0^{2}0^{4}0} \end{array}$$

When you change $\frac{1}{3}$ to a decimal, the calculation goes on forever.

0.3333… is called a **recurring** decimal.

$$\begin{array}{r} 0.3\ 3\ 3\ 3\ 3\ \ldots \\ 3\overline{)1.0^{1}0^{1}0^{1}0^{1}0\ \ldots} \end{array}$$

E1 Change $\frac{2}{3}$ to a recurring decimal.

E2 (a) Work out $\frac{1}{9}$ as a recurring decimal.
(b) Repeat for $\frac{2}{9}$, $\frac{3}{9}$, … and so on.

E3 (a) $\frac{1}{6}$ is half of $\frac{1}{3}$, or $\frac{1}{3}$ divided by 2.
Starting from the recurring decimal for $\frac{1}{3}$, work out the recurring decimal for $\frac{1}{6}$.
(b) Use the result for $\frac{1}{6}$ to find the recurring decimal for $\frac{1}{12}$.

E4 When you work out $\frac{1}{7}$ as a recurring decimal, a whole group of figures recurs:

$$\begin{array}{r} 0.\overbrace{1\ 4\ 2\ 8\ 5\ 7}\ \overbrace{1\ 4\ 2\ 8\ 5\ 7}\ 1\ \ldots \\ 7\overline{)1.0^{3}0^{2}0^{6}0^{4}0^{5}0^{1}0^{3}0^{2}0^{6}0^{4}0^{5}0^{1}0\ \ldots} \end{array}$$

Work out the recurring decimals for $\frac{2}{7}$, $\frac{3}{7}$, … What do you notice?

E5 Find the recurring decimal for
(a) $\frac{1}{13}$ (b) $\frac{1}{17}$

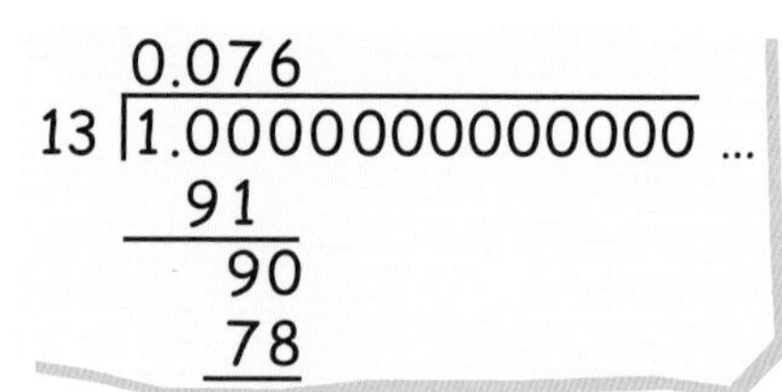

Investigation

- Add some more fractions to each of these lists.

Fractions leading to recurring decimals	$\frac{1}{3}$	$\frac{1}{6}$	$\frac{1}{7}$	$\frac{1}{9}$	$\frac{1}{12}$		
Fractions leading to terminating decimals	$\frac{1}{2}$	$\frac{1}{4}$	$\frac{1}{5}$	$\frac{1}{8}$	$\frac{1}{10}$	$\frac{1}{20}$	$\frac{1}{25}$

- Try to find a rule to tell you when a fraction will lead to a recurring decimal.
- Test your rule on these fractions.

$\frac{1}{21}$ $\frac{1}{30}$ $\frac{1}{11}$ $\frac{1}{15}$ $\frac{1}{40}$ $\frac{1}{16}$ $\frac{1}{80}$ $\frac{1}{36}$ $\frac{1}{52}$ $\frac{1}{64}$

F Mixed questions

F1 During one term, Prakesh had 14 music lessons of $\frac{1}{4}$ hour each, 9 lessons of $\frac{1}{2}$ hour and 7 lessons of $\frac{3}{4}$ hour.

How many hours was this altogether?

F2 Carol cuts a cake into three pieces. She gives $\frac{1}{3}$ to Debbie, $\frac{1}{4}$ to Elly and keeps the rest for herself.

(a) What fraction of the cake does Carol get?

(b) What is the difference between Carol's fraction and Debbie's?

(c) What is the difference between Carol's fraction and Elly's?

F3 Gail is weaving rugs, which are all identical.
In $\frac{1}{4}$ hour she weaves $\frac{2}{3}$ of a rug.
How many rugs will she weave in 4 hours, if she works at the same rate?

F4 (a) From the facts that $\frac{1}{3} = 0.3333\ldots$ and $\frac{1}{2} = 0.5$, it follows that $\frac{1}{3} + \frac{1}{2} =$

$$\begin{array}{r} 0.33333\ldots \\ +\,0.5 \\ \hline 0.83333\ldots \\ \hline \end{array}$$

What fraction is equivalent to the recurring decimal 0.83333... ?

(b) Work out $\frac{1}{3} + \frac{1}{4}$ (i) as a fraction (ii) as a recurring decimal

Egyptian fractions

The ancient Egyptians used only 'unit fractions' like $\frac{1}{2}, \frac{1}{3}, \frac{1}{4}, \frac{1}{5}, \ldots$ (but they did allow $\frac{2}{3}$).
They expressed other fractions by adding unit fractions together.

For example, $\frac{5}{6} = \frac{1}{2} + \frac{1}{3}$ $\quad \frac{19}{20} = \frac{1}{2} + \frac{1}{4} + \frac{1}{5}$

(They didn't repeat the same unit fraction, so $\frac{1}{3} + \frac{1}{3} + \frac{1}{5}$, for example, would not be allowed.)

- Find a way to write each of these fractions in the Egyptian way.
  (You may be able to find more than one way for some of them.)

  $\frac{5}{8}$ $\quad \frac{7}{8}$ $\quad \frac{7}{12}$ $\quad \frac{9}{10}$ $\quad \frac{17}{30}$

What progress have you made?

Statement	Evidence
I can multiply a whole number by a fraction.	**1** Work out (a) $15 \times \frac{3}{4}$ (b) $\frac{5}{8} \times 20$
I can divide a whole number by a fraction.	**2** Work out (a) $12 \div \frac{1}{6}$ (b) $8 \div \frac{2}{3}$
I can change fractions to decimals, including recurring decimals.	**3** Change each of these to a decimal. (a) $\frac{7}{20}$ (b) $\frac{1}{11}$ (c) $\frac{7}{11}$

Review 1

1 The diagram shows a canal boat about to go through a lock.

The sequence of events is this.

- Shutters at the foot of gate A are opened and the water level in the lock falls fast at first but then more slowly, until it is the same as the level of the boat. Then the shutters are closed.
- Gate A is opened, the boat moves forward into the lock, and then gate A is closed.
- Shutters at the foot of gate B are opened and the water level in the lock rises quickly at first but then more slowly until it is the same as the level outside gate B.
- Gate B is opened and the boat moves out of the lock.

Sketch a graph showing the depth of water in the lock during the process.

2 What is the scale factor of the enlargement from

(a) A to B (b) B to A

(c) A to D (d) D to B

(e) C to D (f) A to C

(g) B to D (h) B to C

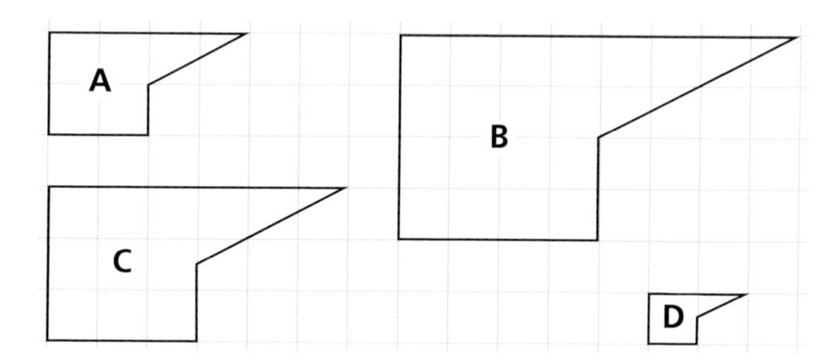

3 Solve these equations.

(a) $7x + 2 = 4x + 17$ (b) $4y + 27 = 9y + 2$ (c) $1.8z + 4.6 = 1.4z + 9.0$

4 The red lines show a way of slicing the triangular prism.
The resulting cross-section has this shape:

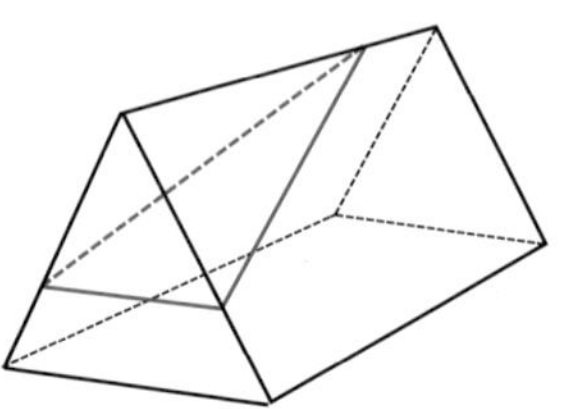

On a sketch of the prism show how to get each of these cross-sections.

(a)  (b)  (c) 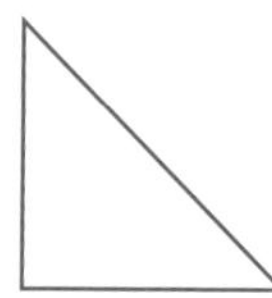

5 Stephen recorded the temperature at midday at his home every day for a year.
He worked out the mean midday temperature for each month and drew this graph.

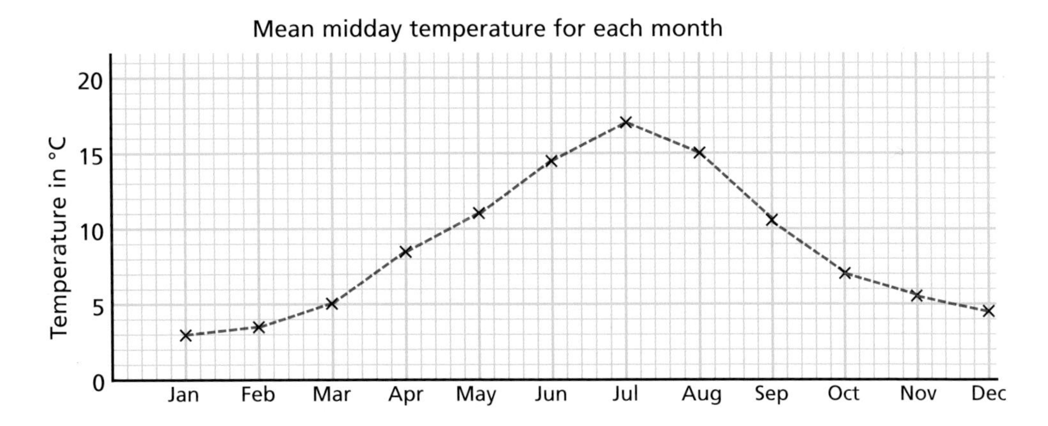

Which of the questions below can be answered from the graph?
Give the answer where possible.

(a) For which month was the mean midday temperature highest?

(b) In which month was the day with the lowest midday temperature?

(c) If the months of July and August are combined, what was the mean midday temperature for the period from July 1 to August 31?

6 A white line on a road is 10 cm wide and 75 metres long.

(a) What is the area of the white line in (i) cm^2 (ii) m^2

(b) One litre of white paint will cover $5\,\text{m}^2$.
How many litres of paint are needed to paint the line?

7 Simplify the following expressions.

(a) $2a + 7 - a - 2$ (b) $5b - 3 - 2b + 6$ (c) $8 - 3c - 6 - c$

(d) $4d - 3 - 7d + 9$ (e) $10 - 4e + e - 3$ (f) $^{-}2f + 7 + 3f - 10$

8 (a) Two identical cubes are placed one on top of the other like this.
How many planes of symmetry does the combined solid have?

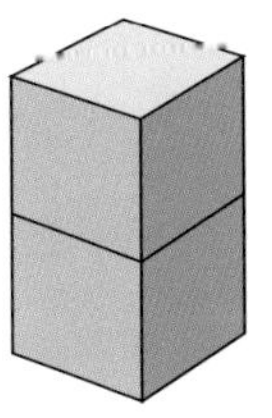

(b) The top cube is rotated through 45°.
How many planes of symmetry does the solid have now?

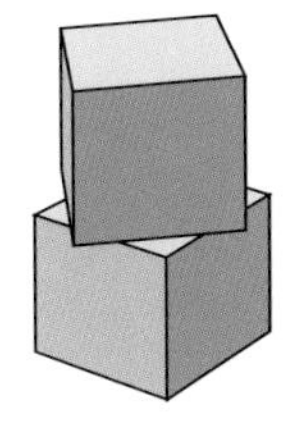

9 Rajesh has a jug which holds $\frac{3}{4}$ litre.

(a) How many jugfuls are needed to fill a tank of capacity 12 litres?

(b) Which of the following calculations is needed to answer part (a)?

A $12 \div \frac{3}{4}$ B $12 \times \frac{3}{4}$ C $\frac{3}{4} \times 12$ D $\frac{3}{4} \div 12$

10 Solve these equations.

(a) $4x - 2 = x + 16$ (b) $6y - 7 = 9y + 2$ (c) $3 + 5z = 9z - 17$

(d) $2(x + 4) = 5(x - 2)$ (e) $3(2 + x) = 7(x - 4)$ (f) $4(1 - x) = 5(x + 8)$

11 These three maps show the same lake.

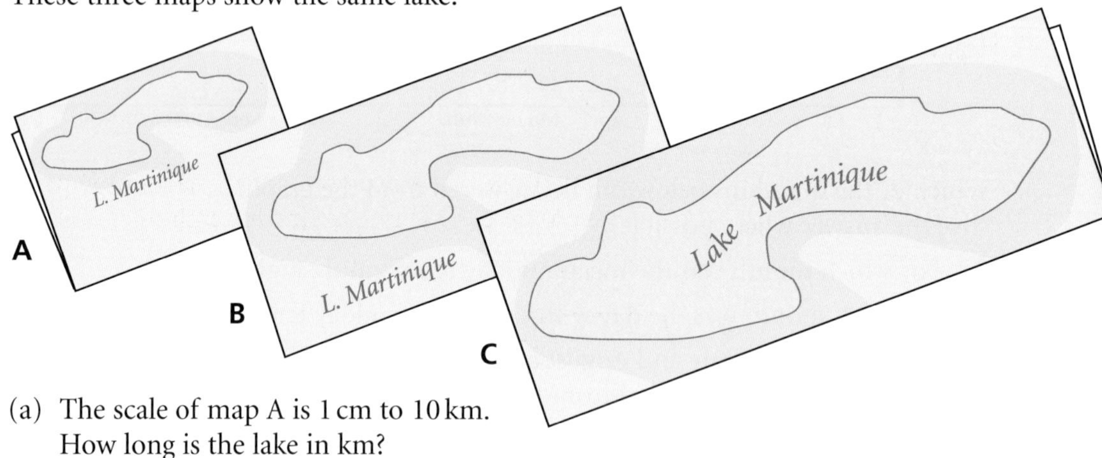

(a) The scale of map A is 1 cm to 10 km.
How long is the lake in km?

(b) What is the scale of map B?
Write it in the form 1 cm to … km.

(c) What is the scale of map C, in the same form?

12 Work out $\frac{4}{13}$ as a recurring decimal.

13 Peter and Grant think of the same number.
Peter doubles his number and subtracts 13.
Grant subtracts 4 from his number and multiplies by 3.

They both get the same answer.
What number did they both think of?

14 As a recurring decimal, $\frac{1}{3} = 0.33333\ldots$

(a) Write down the recurring decimal for $\frac{1}{30}$.

(b) Add the recurring decimals for $\frac{1}{3}$ and $\frac{1}{30}$.

(c) What single fraction gives (b) as a recurring decimal?

(d) Subtract the recurring decimal for $\frac{1}{30}$ from 0.33333…
Write the result as a fraction in its simplest form.

(e) Show that doing $\frac{1}{3} - \frac{1}{30}$ gives the same result as in (d).

9 Transformations

This work will help you

- understand the effects of reflection, rotation and translation
- enlarge a shape from a centre
- understand combinations of transformations
- use a computer to investigate transformations

A Reflection

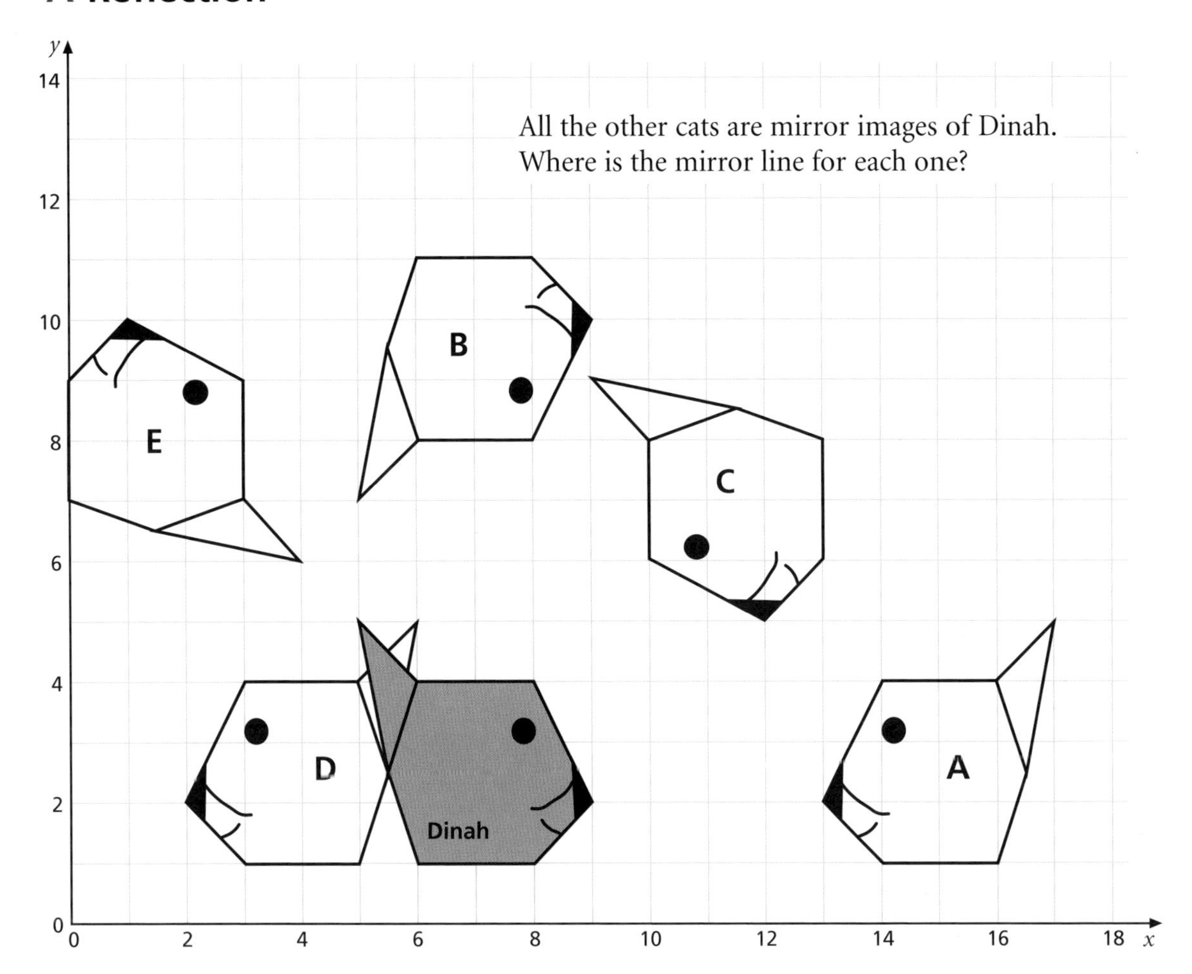

A1 This question is on sheet 193.

B Lines on a grid

On a grid, it is easy to give the positions of points – just use coordinates.

We also need to be able to describe lines, using equations.

Find the line marked *a*.

Here are some of the points on this line.

$(2, 0)$ $(2, 3)$ $(2, ^-4)$ $(2, ^-1)$

The *x*-coordinate of every point on line *a* is 2.

So the equation of line *a* is $\mathbf{x = 2}$.

Find the line marked *b*.

The *y*-coordinate of every point on line *b* is 3, so the equation of line *b* is $\mathbf{y = 3}$.

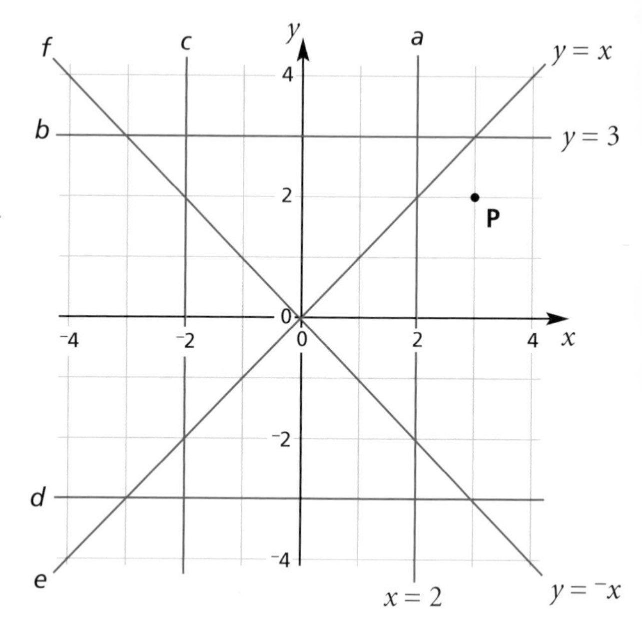

B1 Write down the equation of (a) line *c* (b) line *d*

B2 (a) Explain why the equation of line *e* is $y = x$.

(b) Explain why the equation of line *f* is $y = {}^-x$.

B3 (a) What is another name for the line whose equation is $x = 0$?

(a) What is another name for the line whose equation is $y = 0$?

B4 The point P has coordinates $(3, 2)$.

(a) P is reflected in the line $x = 2$. What are the coordinates of the reflection?

(b) P is reflected in the line $y = 1$. What are the coordinates of the reflection?

(c) P is reflected in the line $y = x$. What are the coordinates of the reflection?

(d) P is reflected in the line $y = {}^-x$. What are the coordinates of the reflection?

B5 Start with point P.
Reflect it in $x = 1$. Then reflect the result in $y = 3$.
Then reflect the result in $y = {}^-x$. Finally reflect the result in $y = x$.
What are the coordinates of the point you finish at?

B6 Start with P. Reflect in $y = {}^-1$, then $y = {}^-x$, then the *y*-axis.
Where do you finish?

C Rotation

The diagram below is on sheet 194.

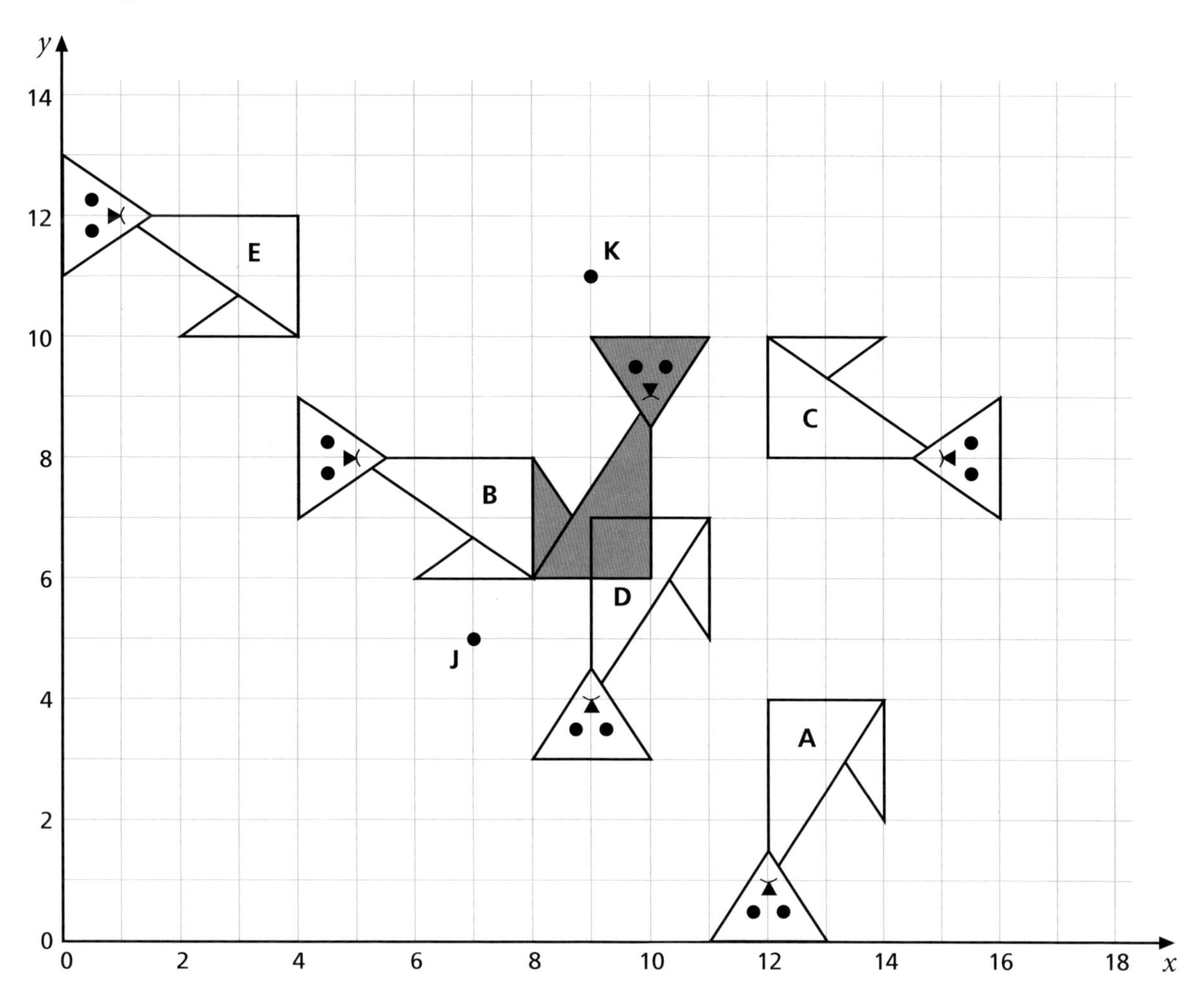

Rotate the grey cat

- a half turn (180°) about point J (7, 5)
- a quarter turn (90°) clockwise about K (9, 11)
- a quarter turn anticlockwise about K

Can you describe how all the other rotations were done?

C1 This question is on sheet 195.

C2 This question is on sheet 196.

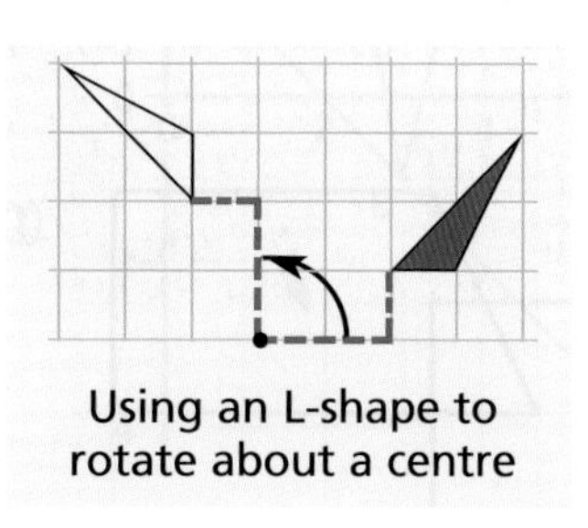

Using an L-shape to rotate about a centre

H Combining transformations

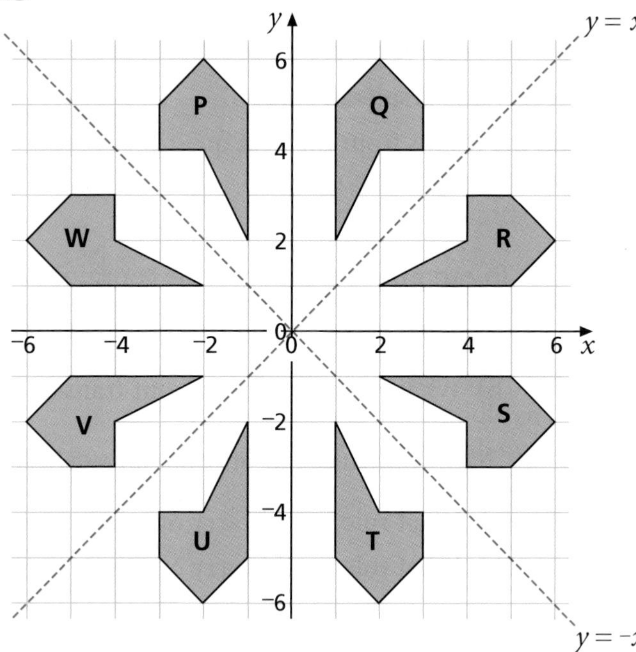

H1 (a) Which shape is the image of P after reflection in the y-axis then reflection in the x-axis?

(b) Describe the single transformation which has the same effect as these two.

H2 What are the missing entries in the table below?

Object	First transformation	Second transformation	Final image	Single equivalent transformation
R	Rotation about (0, 0) 90° clockwise	Reflection in $y = x$	(a)	(b)
V	Reflection in $y = {}^{-}x$	Rotation about (0, 0) 180°	(c)	(d)
Q	Rotation about (1, 2) 90° clockwise	Translation $\begin{bmatrix} 1 \\ -3 \end{bmatrix}$	(e)	(f)
W	Reflection in the y-axis	(g)	Q	(h)
P	(i)	Rotation about (0, 0) 90° clockwise	S	(j)
Q	Rotation about (0, 0) 180°	(k)	(l)	Reflection in the x-axis
(m)	Rotation about (0, 0) 90° anticlockwise	Reflection in $y = x$	S	(n)
(o)	Reflection in the x-axis	(p)	R	Rotation about (0, 0) 90° clockwise
(q)	(r)	Rotation about (0, 0) 180°	R	Reflection in the y-axis

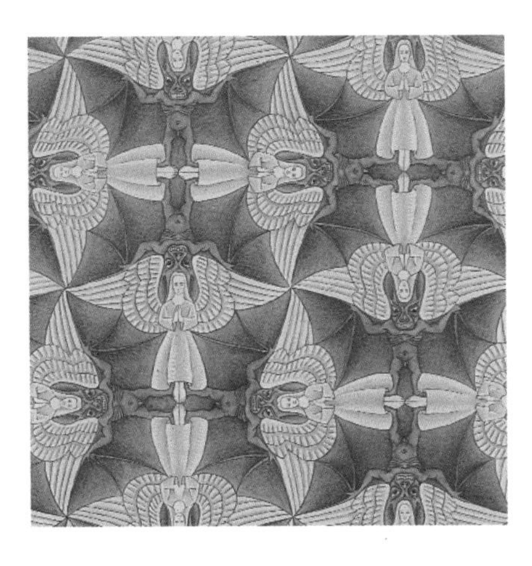

What transformations connect the shapes in this pattern?

Artist: M C Escher
Title: Angels and Devils

What progress have you made?

Statement

I can reflect and rotate shapes accurately.

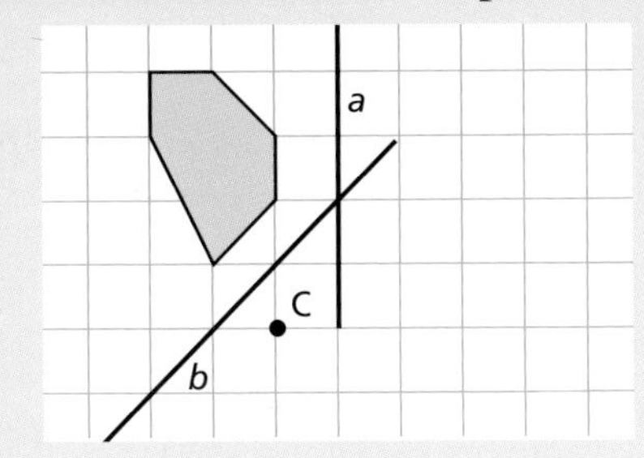

Evidence

1 Copy this diagram on to squared paper.

(a) Reflect the shaded shape in line a.
Label the image (a).

(b) Reflect the shaded shape in line b.
Label the image (b).

(c) Rotate the shaded shape through 90° anticlockwise about C.
Label the image (c).

I can describe a transformation.

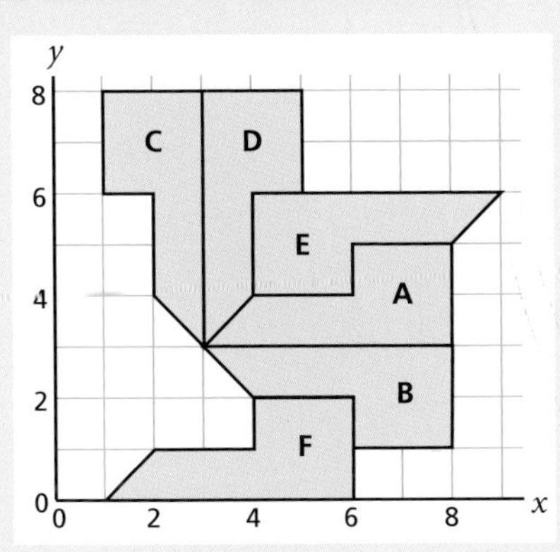

2 Describe the single transformation which maps

(a) A on to B
(b) A on to C
(c) A on to D
(d) A on to E
(e) A on to F
(f) E on to F

I can describe the effect of combining two transformations.

3 Shape B is rotated 90° anticlockwise about (3, 3) and the result is reflected in $x = 3$. What single transformation has the same effect as these two?

10 True, iffy, false 1

For group and class discussion

Some of the statements below are always true.
Some of them are false (never true).
Some of them are 'iffy' (true if ...).

For example: **A multiple of 5 ends in a 5.**

This statement is iffy because it is sometimes true (5, 15, 25, 35, ...) and sometimes false (10, 20, 30, 40, ...).

Look at each statement.
Decide whether it is always true, or iffy, or false.
If it is iffy, try to modify it so that it becomes true.

For example: **A multiple of 5 ends in 5 or 0.**

or **An odd multiple of 5 ends in 5.**

Don't forget!
'Numbers' includes fractions, decimals and negative numbers.

1 To multiply by 10, add an extra 0.

2 An even number is divisible by 2.

3 Finding two fifths of a number is the same as dividing it by 5 then multiplying by 2.

4 Squaring a number gives a higher number.

5 You divide a number by 2 and then by 10.
The answer is different if you divide it by 10 then by 2.

6 The total of two numbers is greater than their difference.

7 If the total of two numbers is odd, so is their difference.

8 A prime number is an odd number.

9 You double a number and add 10.
The answer is different if you add 10 first and then double.

10 Adding two negative numbers gives a positive number.

Linear equations and graphs

This work will help you
- understand how gradient is measured
- describe the graph of an equation of the form $y = mx + c$
- find the equation of a given straight line graph

A Gradient

The introductory activity is described in the teacher's guide.

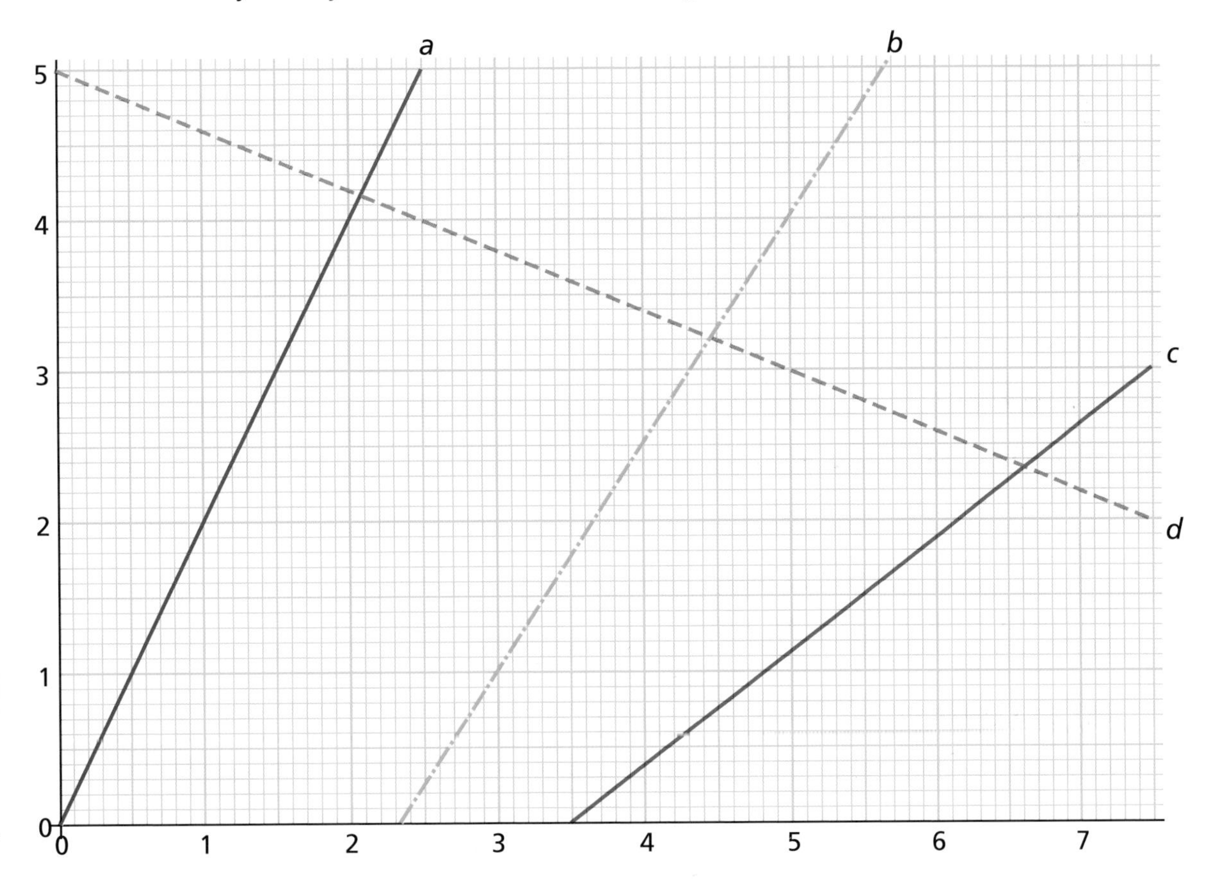

- How could you measure the steepness of a line?

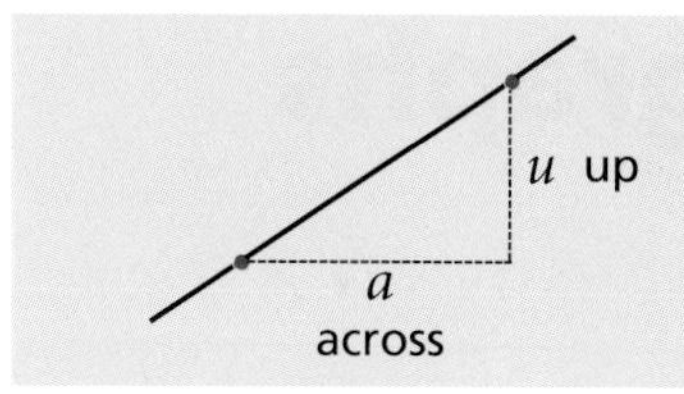

$\frac{u}{a}$ is called the **gradient** of the line.

Railway gradients

If a hill on a railway line is too steep, the engine's wheels will slip on the rails. The steepest gradient an ordinary engine can cope with is 0.09.

Mountain railways sometimes use the 'rack and pinion' system. A toothed wheel (the pinion) on the engine runs along a toothed rail (the rack).

By this method the gradient can go up to 0.125.
If there is only a single car, the gradient can be up to 0.5.

In a funicular railway, the car is fixed to a cable which pulls it up the hill.
The steepest funicular railway has a gradient of 0.89.

Type of railway	Ordinary	Rack and pinion (train)	Rack and pinion (single car)	Funicular
Maximum gradient	0.09	0.125	0.5	0.89

A1 Find the gradient of each of these railways, to two decimal places.
Write down a type of railway which would be suitable for each one.

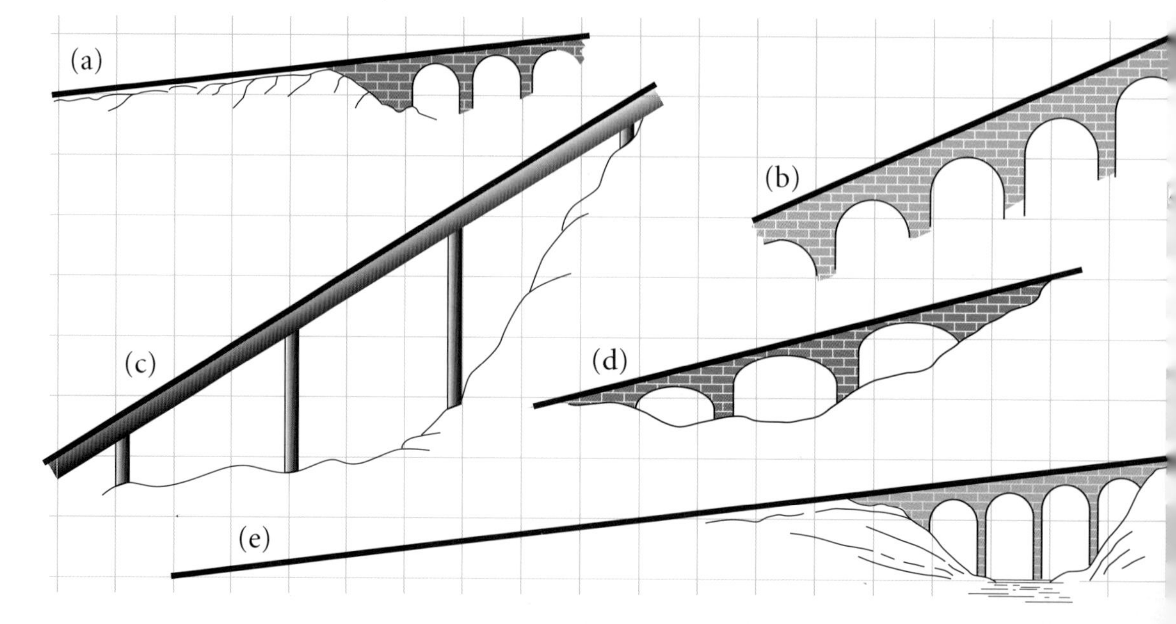

B Investigating linear graphs

This work is best done on a computer, using a graph drawing program.

- Draw the graph of $y = 2x$, for values of x from $^-3$ to 3.

x	$^-3$	$^-2$	$^-1$	0	1	2	3
y	$^-6$	$^-4$	$^-2$	0	2	4	6

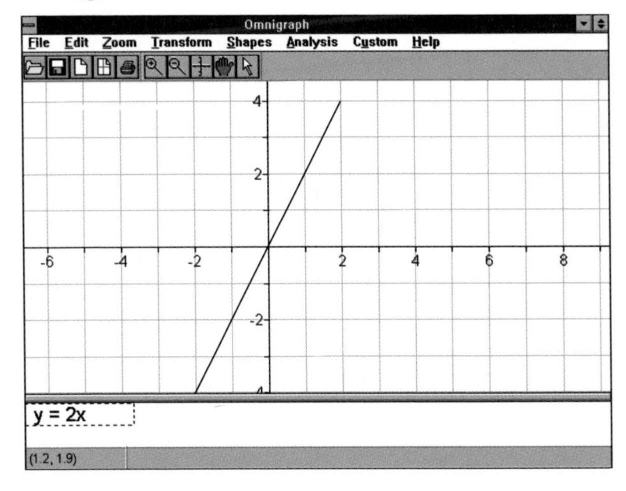

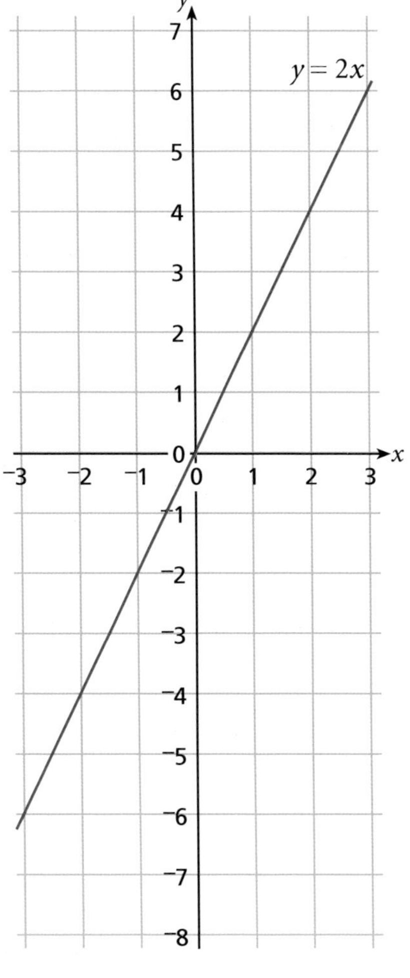

What is the gradient of the line $y = 2x$?

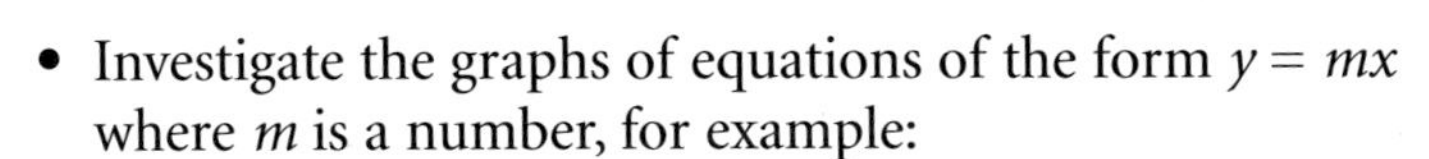

- Investigate the graphs of equations of the form $y = mx$ where m is a number, for example:

  $y = 3x \qquad y = 1.5x \qquad y = \frac{1}{2}x$

  What is the gradient of each line?

- Investigate the graphs of equations of the form $y = 2x + c$ where c is a number (which may be negative), for example:

  $y = 2x + 1 \qquad y = 2x + 5 \qquad y = 2x - 3$

  What can you say about the gradients of these graphs?

- Investigate the graphs of equations of the form $y = mx + 2$ where m is a number, for example:

  $y = 2x + 2 \qquad y = 3x + 2 \qquad y = \frac{1}{2}x + 2$

  What do all these graphs have in common?

- Can you predict, before drawing, what the gradient of each of these lines will be, and where the line will cross the y-axis?

  $y = 3x - 2 \qquad y = 2x + 3 \qquad y = \frac{1}{2}x + 1$

- Extend the investigation to include equations like $y = {}^-3x + 2$, $y = {}^-2x - 3$, …

C Gradient and intercept

These equations

$y = 3x + 5 \qquad y = 5x \qquad y = \frac{1}{2}x - 2$

are all examples of the form

$$y = mx + c$$

where m and c are numbers.

The graphs of equations of the form $y = mx + c$ are straight lines.

The number m is the gradient of the line.

The number c is the **intercept** on the y-axis.

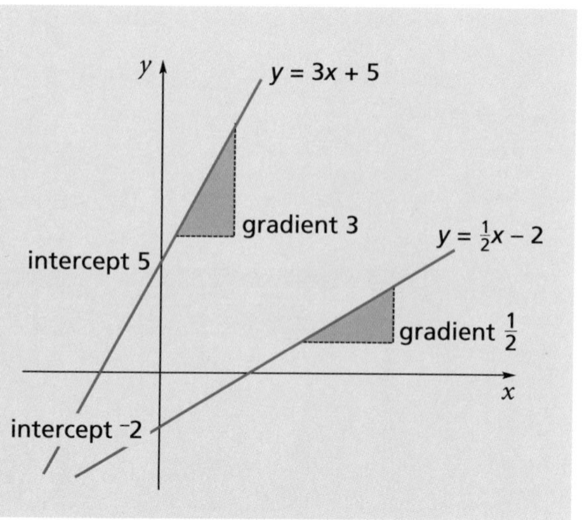

C1 Write down (i) the gradient (ii) the intercept on the y-axis of each of these linear graphs.

(a) $y = 7x + 2$ (b) $y = 0.2x - 4$ (c) $y = 8x$ (d) $y = x - 7$ (e) $y = 5$

C2 Write down the equation of each of these linear graphs.

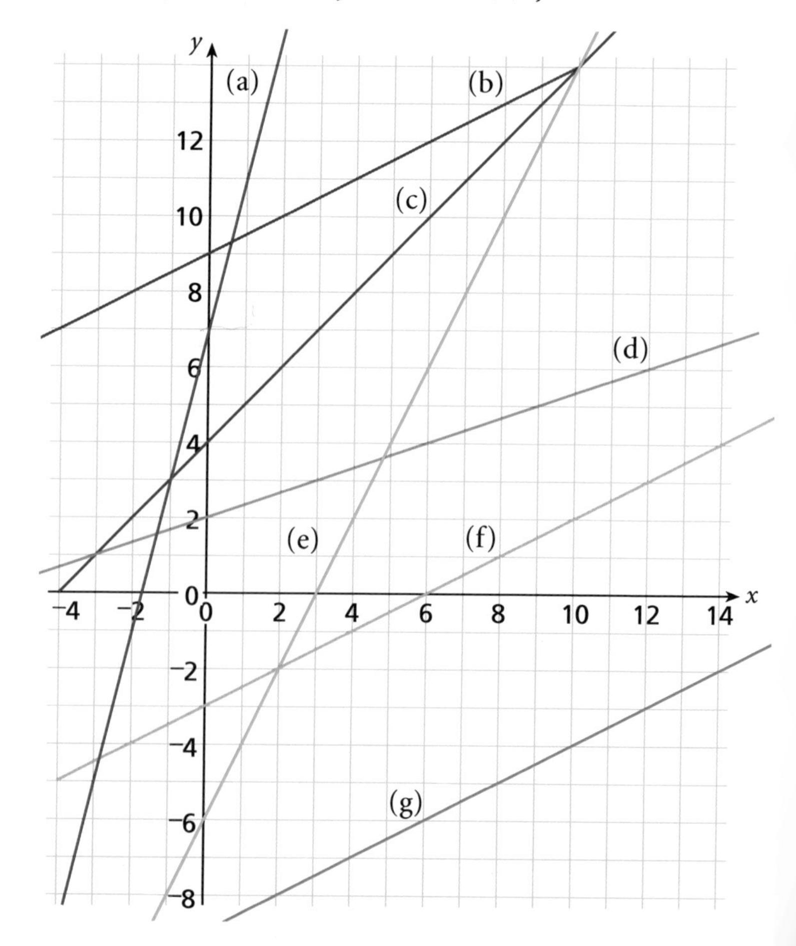

D Negative gradients

Here is the graph of $y = {}^-2x + 7$.

Its gradient $= \dfrac{\text{distance up}}{\text{distance across}} = \dfrac{^-4}{2} = {}^-\mathbf{2}$

Lines with negative gradients slope downwards.

The equation $y = {}^-2x + 7$ can also be written as

$y = 7 - 2x$

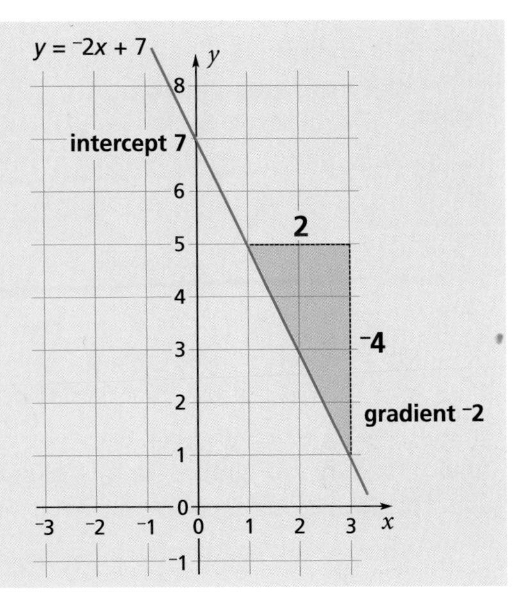

D1 Write down (i) the gradient (ii) the intercept on the y-axis of each of these linear graphs.

(a) $y = {}^-4x + 1$ (b) $y = {}^-\frac{1}{2}x - 6$ (c) $y = {}^-1.5x$ (d) $y = 6 - 2x$ (e) $y = 7 - x$

D2 Find the equation of each of these lines.

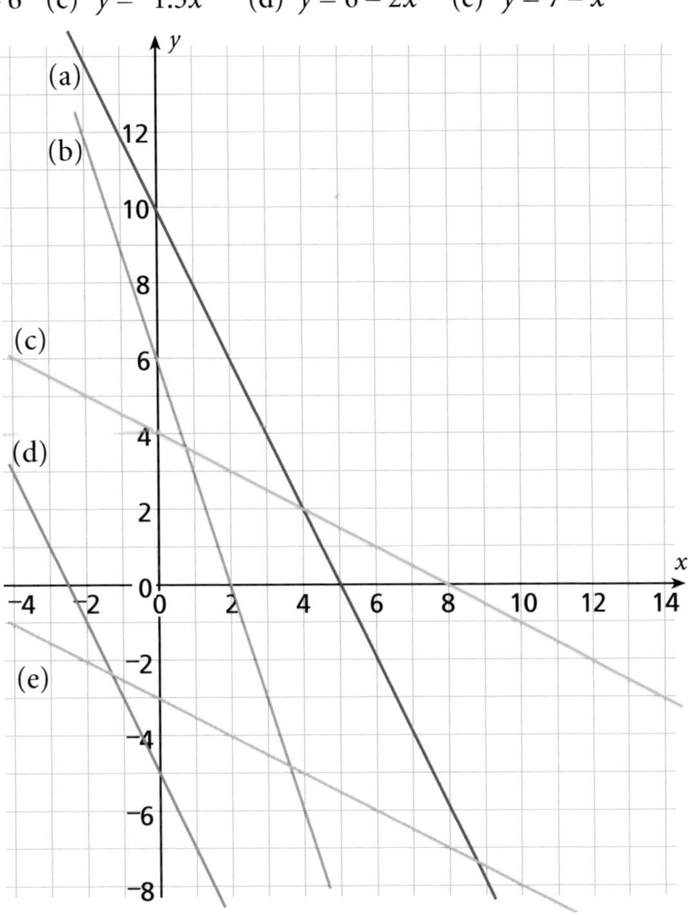

D3 You need sheet 198.

D4 You need sheet 199.

D5 On this diagram, the graph of $y = {}^-x - 2$ has been drawn.
There are two other graphs shown, A and B.
Which equations below fit A?
Which equations fit B?
(Hint: there is more than one in each case.)

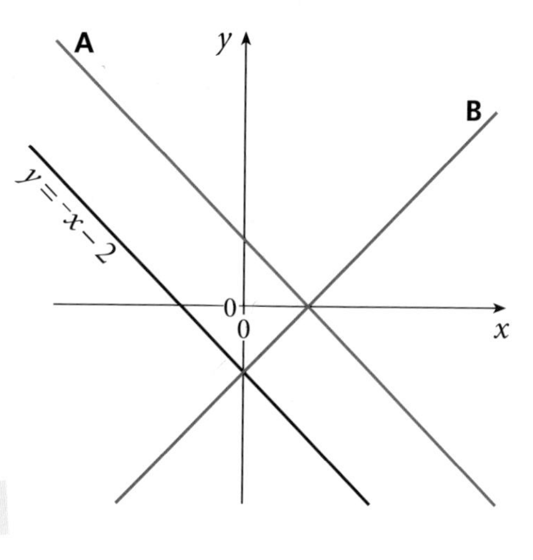

$y = {}^-x + 2$ $y = 2x - 2$

$y = x + 2$ $y = x + {}^-2$ $y = x - 2$

$y = {}^-x - 2$ $y = 2 - x$ $y = {}^-2x - 2$

D6 There are four graphs on this diagram.
There are four equations below.
Which graph goes with which equation?

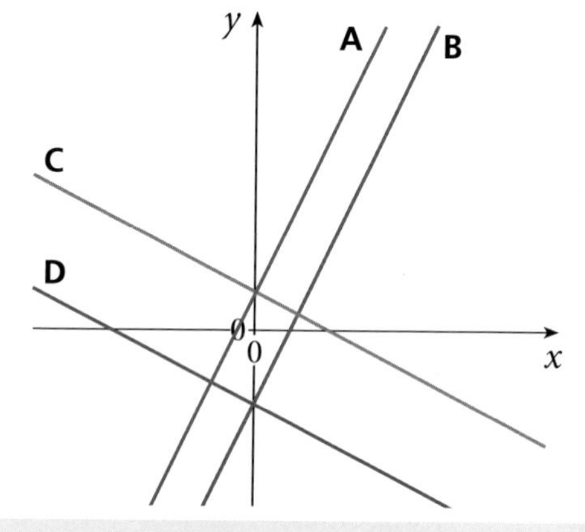

$y = 2x - 2$ $y = {}^-\frac{1}{2}x - 2$

$y = 1 - \frac{1}{2}x$ $y = 2x + 1$

What progress have you made?

Statement	Evidence
I can work out the gradient of a line.	**1** Find the gradient of each of these lines.
I can find the gradient and the intercept on the y-axis for a graph whose equation is in the form $y = mx + c$.	**2** Find (i) the gradient (ii) the intercept on the y-axis of the line whose equation is (a) $y = 3x - 4$ (b) $y = 4 - 2x$
I can find the equation of a straight-line graph.	**3** Find the equation of each of these lines.

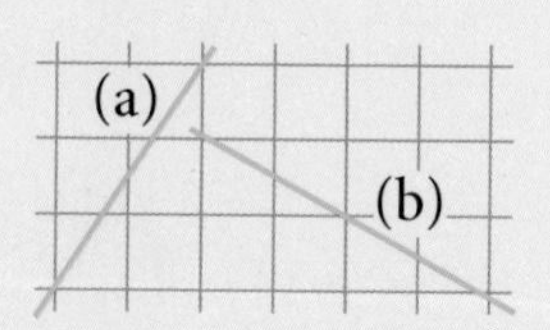

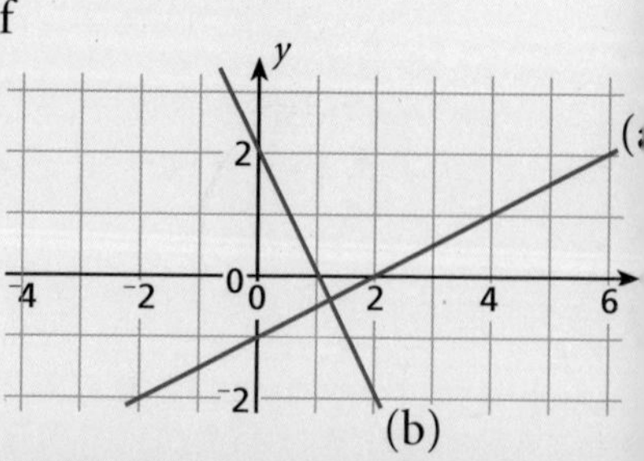

Percentage changes

This is about percentage increases and decreases.
The work will help you

- calculate the result of a percentage increase or decrease
- express increases or decreases as percentages

A Percentages and decimals

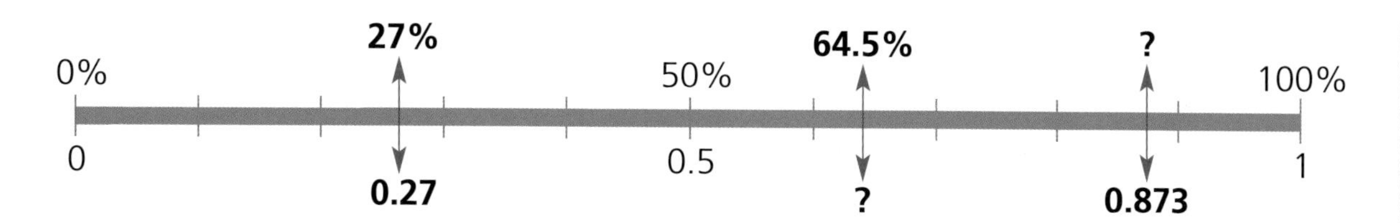

A1 Write each of these percentages as a decimal.

(a) 30% (b) 32% (c) 32.5% (d) 39.6% (e) 40.3%

A2 Write each of these decimals as a percentage.

(a) 0.7 (b) 0.74 (c) 0.748 (d) 0.406 (e) 0.333

A3 Write each of these percentages as a decimal.

(a) 6% (b) 6.3% (c) 4.5% (d) 1.9% (e) 0.7%

A4 Write each of these decimals as a percentage.

(a) 0.02 (b) 0.026 (c) 0.091 (d) 0.005 (e) 0.1575

A5 Calculate each of these to the nearest penny.

(a) 35% of £240 (b) 35.3% of £240 (c) 67.5% of £450 (d) 8.5% of £320
(e) 60.3% of £210 (f) 7.5% of £660 (g) 10.8% of £290 (h) 0.6% of £880

A6 Change $\frac{1}{8}$ to (a) a decimal (b) a percentage

A7 Change $\frac{5}{8}$ to (a) a decimal (b) a percentage

A8 In a class of 32 children there are 22 boys and 10 girls.
What percentage of the class are boys?

A9 57 students took an examination. 34 of them passed.
What percentage of the students passed? Give your answer to the nearest 0.1%.

B Percentage increases

Every so often bus and train fares go up.
The increase is usually given as a percentage.

If fares go up by 25%, this means 25% of the old fare is added on to make the new fare.

Fares up by 25%

Commuters will be hit again in the new year as the train companies raise fares by 25%. Promises of better punctuality, greater

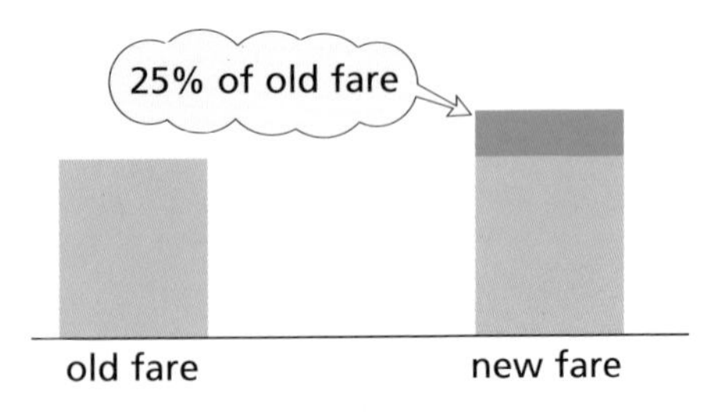

$\frac{25}{100}$ of the old fare is added on.

So the new fare is $1\frac{25}{100}$ times the old fare.

This can also be written

old fare × **1.25** = new fare

To increase by 25%, you multiply by ***1.25***

B1 Calculate the new fare after a 25% increase when the old fare is

(a) £40 (b) £4 (c) £60 (d) £42 (e) £1.80

B2 If fares go up by 15%, what number do you multiply by to get the new fare?

B3 Calculate the new fare after a 15% increase when the old fare is

(a) £30 (b) £8 (c) £50 (d) £76 (e) £1.20

B4 Fares go up by 12%. What is the new fare if the old fare is

(a) £50 (b) £85 (c) £6.50

B5 Fares go up by 8%. Karl says you have to multiply old fares by 1.8.
Is he right? If not, what do you multiply by?

B6 Calculate the new fare after an 8% increase when the old fare is

(a) £50 (b) £7 (c) £20 (d) £36 (e) £1.50

B7 Fares go up by 4.5%.

(a) What number do you multiply by to get the new fare?

(b) Calculate, to the nearest penny, the new fare if the old fare is £18.

B8 Calculate, to the nearest penny, the new fare in each case below.

	Old fare	Percentage increase
(a)	£44	3.5%
(b)	£68	4.2%
(c)	£120	0.9%
(d)	£37.50	13.5%

C Percentage decreases

In a sale, the price reduction is often given as a percentage.

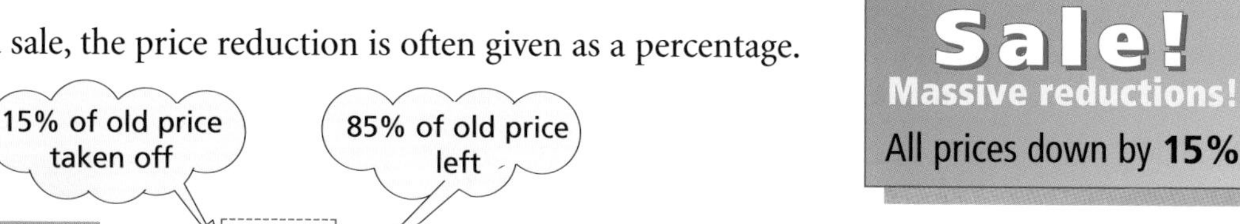

old price new price

$\frac{15}{100}$ of the old price is taken off. That leaves $\frac{85}{100}$ of the old price.

So the new price is $\frac{85}{100}$, or **0.85**, of the old price.

This can also be written

old price × **0.85** = new price

To decrease by 15%, you multiply by ***0.85***

C1 Reduce each of these prices by 15%.

(a) £20 (b) £38 (c) £72 (d) £68.40 (e) £49.60

C2 What number do you multiply by to reduce prices by 35%?

C3 Reduce each of these prices by 35%.

(a) £30 (b) £76 (c) £180 (d) £98.60 (e) £247.20

C4 What number do you multiply by to reduce prices by 6%?

C5 Reduce each of these prices by 6%.

(a) £44 (b) £28 (c) £610 (d) £380 (e) £1470

C6 What do you multiply by in order to reduce prices by

(a) 3% (b) 9% (c) 11% (d) 30% (e) 14.5%

C7 Reduce

(a) £240 by 8% (b) £750 by 4% (c) £60 by 11% (d) £80 by 13.6%

C8 (a) Increase £45 by 5%. (b) Reduce £65 by 12%.

(c) Increase £94 by 7%. (d) Reduce £280 by 16.5%.

C9 Every percentage increase or decrease corresponds to a multiplier, for example:

5% increase: multiply by **1.05** 8% decrease: multiply by **0.92**

What percentage change corresponds to each of these?
Make it clear whether it is an increase or a decrease.

(a) Multiply by 1.17 (b) Multiply by 1.02 (c) Multiply by 0.94

(d) Multiply by 0.88 (e) Multiply by 1.4 (f) Multiply by 0.7

(g) Multiply by 1.725 (h) Multiply by 0.65 (i) Multiply by 0.235

D Successive increases and decreases

D1 The population of a new town is planned to increase by 15% during this year. Then it is planned to increase by 8% during next year.

This diagram shows what will happen.

× 1.15 → × 1.08 →

population at start of this year | population at end of this year | population at end of next year

The population at the start of this year was 20 000.
Calculate the planned population at the end of (a) this year (b) next year

D2 A car manufacturer increases prices by 5% on 1 April.
On 1 October it increases prices by 10%.
Calculate the price on 2 October of a car whose price in March is

(a) £5600 (b) £7200 (c) £10 500 (d) £12 550 (e) £18 000

D3 Does it make any difference to the final price if the prices in D2 increase first by 10% and then by 5%? Explain your answer.

D4 A shop increases its prices by 20% in September.
It then reduces prices by 5% for a sale in December.

(a) What number is missing in this diagram?

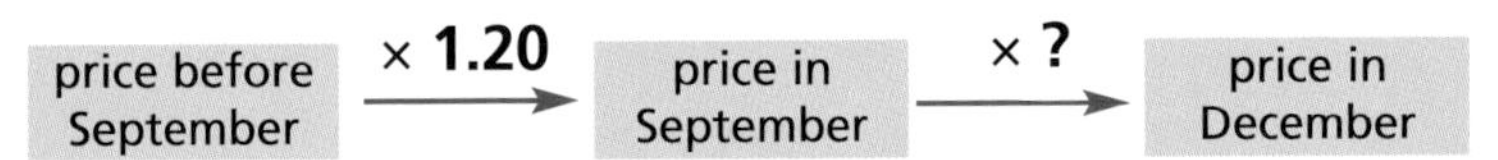

(b) Calculate the December price of an article whose price before September was

(i) £60 (ii) £55 (iii) £8.80 (iv) £18.50 (v) £405

(c) Does it make any difference to the December price if the shop reduces prices by 5% in October and then increases prices by 20% in December? Explain.

D5 Calculate the final prices in this table. Round to the nearest penny if necessary.

	Starting price	First percentage change	Second percentage change	Final price
(a)	£80	up by 15%	up by 30%	
(b)	£140	down by 20%	down by 5%	
(c)	£350	down by 6%	up by 45%	
(d)	£8.20	up by 32%	down by 18%	
(e)	£8.20	down by 18.5%	up by 32.5%	

D6 The number of fish in Karl's pond goes up by 10% in the first year and then by 10% in the second year. Karl says that in two years the population must go up by 20%. Is he right? If not, why not?

E Expressing changes as percentages

Example 1

A fare goes up from £35 to £37.10. What is the percentage increase?

You need to find the multiplier from 35 to 37.10.

35 —× ?→ 37.10

To find the multiplier, **divide** 37.10 by 35. → 37.10 ÷ 35 = 1.06

The multiplier 1.06 corresponds to a **6% increase**. → The percentage increase is 6%.

E1 The price of a TV goes up from £250 to £270.
Calculate the percentage increase.

E2 Calculate the percentage increase in each case below.

(a) From £35 to £39.90 (b) From £28.50 to £35.34 (c) From £66.40 to £76.36
(d) From £85 to £113.90 (e) From £42 to £47.25 (f) From £55 to £61.49

E3 The price of a stereo goes up from £129 to £145.
Calculate the percentage increase, to the nearest 1%.

E4 An air fare goes up from £385 to £405.
Calculate the percentage increase to the nearest 0.1%.

Example 2

A price goes down from £45 to £41.40. What is the percentage decrease?

You need to find the multiplier from 45 to 41.40.

45 —× ?→ 41.40

To find the multiplier, divide 41.40 by 45. → 41.40 ÷ 45 = 0.92

The multiplier 0.92 corresponds to an **8% decrease**. → The percentage decrease is 8%.

E5 Calculate the percentage decrease in each case below.

(a) From £48 to £45.60 (b) From £124 to £105.40 (c) From £72 to £58.32
(d) From £82 to £63.14 (e) From £248 to £136.40 (f) From £155 to £151.90

E6 Calculate the percentage change in each case below.
Say whether it is an increase or a decrease.

(a) From £45 to £48.60 (b) From £125 to £107.50 (c) From £38 to £35.34
(d) From £650 to £429 (e) From £140 to £150.50 (f) From £160 to £141.60

F Problems

F1 (a) Sanjit's weekly pay goes up from £46 to £57.50. What is the percentage increase?

(b) Later, his pay goes down from £57.50 to £46. What percentage drop is this?

F2 David is given £80 on his 14th birthday. He opens a bank account with it.
At then end of every year, the bank increases the amount in the account by 5%.

If David does not take any money out, how much will there be in the account on his 18th birthday?

F3 The Guinea Pigs Club collects money for charity.
Last year there were 350 people in the club and they collected £5600 altogether.
This year there are 490 people and they collected £10 192 altogether.

Calculate each of these.

(a) The percentage increase in the number of people between last year and this year

(b) The percentage increase in the amount collected

(c) The mean amount collected per person last year

(d) The percentage change in the mean amount between last year and this year

*F4 A 20% increase is followed by another 20% increase.
Calculate the overall percentage increase.

*F5 In 1984, the chocolate ration went down by 10% in January.
In April it was increased by 10%.
What was the overall percentage change in the chocolate ration?

*F6 The population of a country increases by 15% every year.
After how many years will the population be double what it is now?

What progress have you made?

Statement	Evidence
I can increase or decrease an amount by a given percentage.	1 (a) Increase £345 by 8%. (b) Decrease £165 by 14%.
I can express an increase or decrease as a percentage.	2 Express each of these as a percentage change. (a) £248 increasing to £287.68 (b) £485 decreasing to £451.05
I can calculate successive percentage changes.	3 A shop increases its prices by 8% on 1 June. On 1 September it reduces prices by 15%. A TV cost £260 before 1 June. How much does it cost after 1 November, to the nearest £?

13 True, iffy, false 2

For group and class discussion

Some of the statements below are always true.
Some of them are false (never true).
Some of them are 'iffy' (true if ...).

Example:

The diagonal of a rectangle is a line of symmetry.

This is an iffy statement because it is false for this rectangle

but true for this one.

(A square is a special kind of rectangle.)

The statement can be modified so that it is true:

The diagonal of a rectangle is a line of symmetry if the rectangle is a square.

Decide whether each statement below is always true, or iffy or false.
If it is iffy, try to modify it so that it becomes true.

1 Some triangles have two obtuse angles.

2 The diagonals of a rectangle cross at right angles.

3 A parallelogram has no lines of symmetry.

4 Four straight lines cross at six points.

5 The longest side of a triangle is shorter than the total of the other two sides.

6 A right-angled triangle has one line of symmetry.

7 If A is 20 cm away from B and C is 10 cm away from B, then A and C are 10 cm apart.

8 If two lines are both at right angles to a third line, then the first two lines are parallel.

9 If two rectangles overlap, the shape of the overlap is also a rectangle.

Probability from experiments

This is about experiments involving chance.
The work will help you
- understand relative frequency
- use relative frequency as an estimate of probability

A Experiments

T

Dropping a spoon

You need a plastic spoon and sheet 200.

When you drop a spoon, it can land the right way up

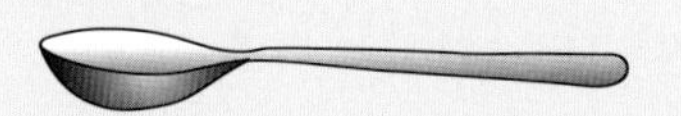

or upside down

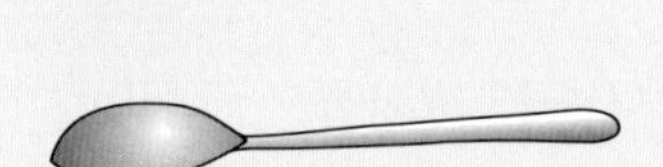

Which way it lands is a matter of chance.

- Which do you think is more likely?
- What do you think is the probability of landing the right way up?
  (You could mark your estimate on a probability scale from 0 to 1.)

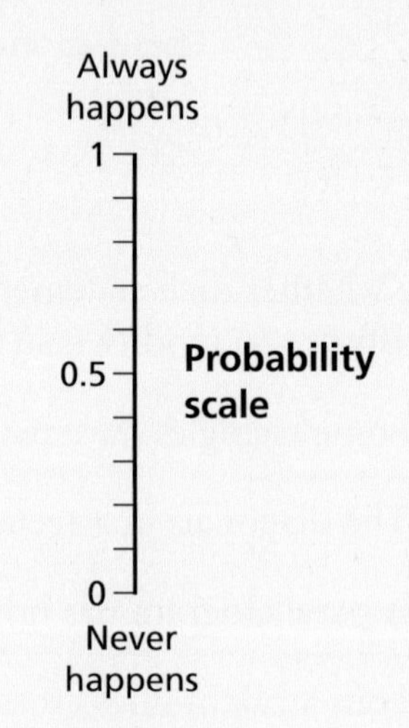

Drop a spoon 10 times and record which way it lands.

Collect together the results for the whole class and find the **relative frequency** of 'right way up' landings.

$$\text{Relative frequency of 'right way up'} = \frac{\text{number of times spoon landed right way up}}{\text{number of times it was dropped}}$$

The results of the experiment can be recorded on sheet 200.

Here is an example:

R means 'right way up',
U 'upside down'.

A 'success' is the spoon landing right way up.

Outcomes										Number of successes	Total number of successes so far	Total number of trials so far	Relative frequency
R	R	U	R	U	U	R	R	R	R	7	7	10	0.7
U	R	U	U	R	R	U	R	R	U	5	12	20	0.6
R	R	R	U	R	U	R	R	U	U	6	18	30	0.6
U	R	R	U	U	R	U	R	U	R	5	23	40	0.58
R	U	R	R	R	U	R	R	R	R	8	31	50	0.62
U	R	U	R	R	R	U	R	U	R	6	37	60	0.62
U	U	R	U	R	U	R	R	R	U	5	42	70	0.6
R	R	U	R	R	R	R	U	U	R	7	49	80	0.61
R	R	R	U	R	R	U	R	R	R	8	57	90	0.63
U	R	U	U	R	R	R	R	U	R	6	63	100	0.63

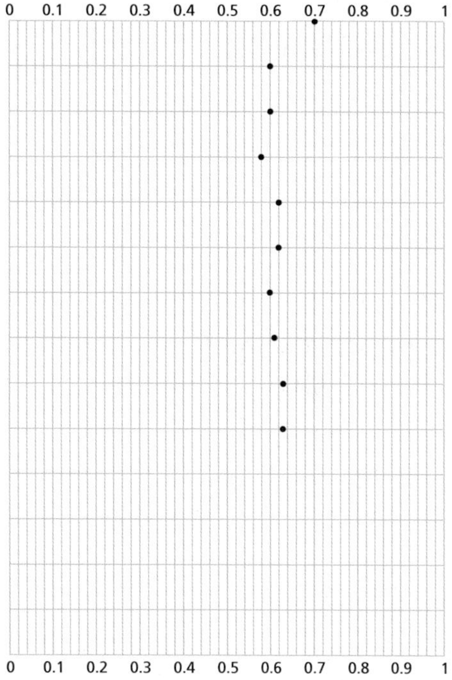

You can use the relative frequency as an **estimate** of the probability that the spoon lands the right way up.

The other way of working out probabilities is based on **equally likely outcomes.**

For example, when a fair dice is rolled, each face is equally likely to come up, so the probability of each face coming up is $\frac{1}{6}$.

With a spoon there are no equally likely outcomes.
You have to use experimental data to estimate the probability.
The more data you have, the better the estimate.

More experiments

Dropping a multilink cube

When you drop a multilink cube, there are four ways it can land.

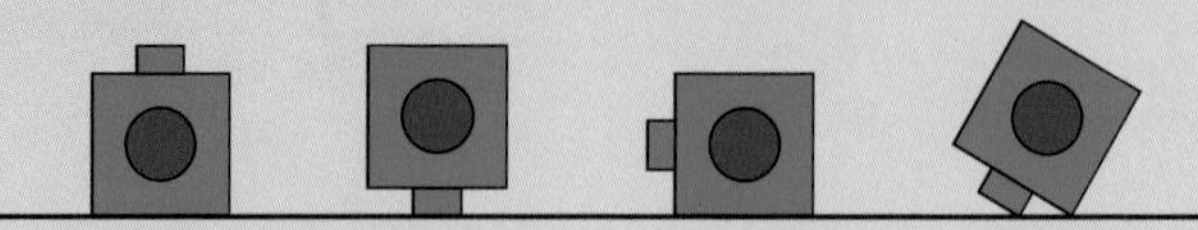

Before doing the experiment, guess the probability of each of these ways of landing.

Now plan and carry out an experiment to estimate the probabilities.

It saves time if you drop ten cubes at a time.

Drawing pin experiment

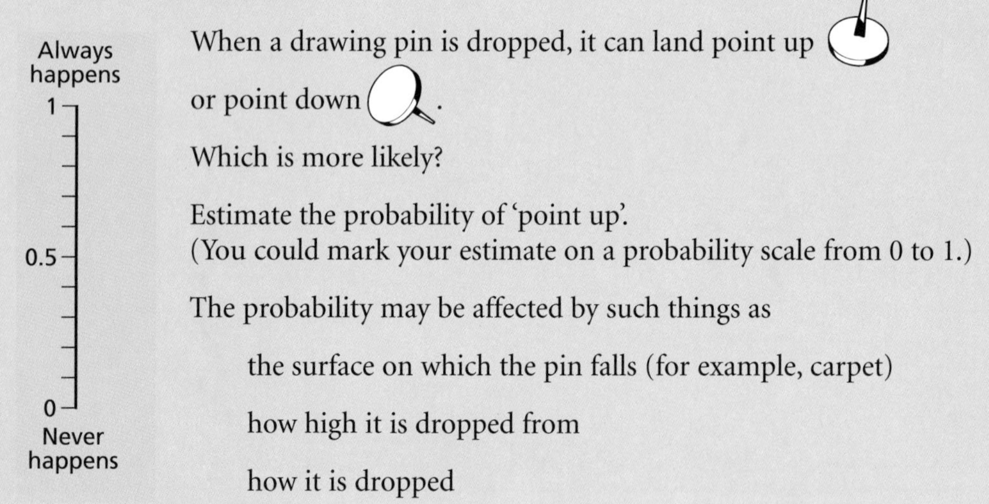

When a drawing pin is dropped, it can land point up or point down.

Which is more likely?

Estimate the probability of 'point up'.
(You could mark your estimate on a probability scale from 0 to 1.)

The probability may be affected by such things as

the surface on which the pin falls (for example, carpet)

how high it is dropped from

how it is dropped

Plan and carry out an experiment to estimate the probability of 'point up'.

You could compare results under different conditions.

Money down the drain

You need sheet 201 and a penny.

Tap the penny on to the 'drain'. The tap only counts if the penny ends up wholly inside the rectangle. If it ends wholly on a white part, it goes 'down the drain'.

Estimate the probability that a penny ending on the drain goes down the drain.

B Relative frequency

Beth and Steve were doing a traffic survey.

They stood by a junction where traffic could turn either left or right.

They noted down which way each vehicle turned.

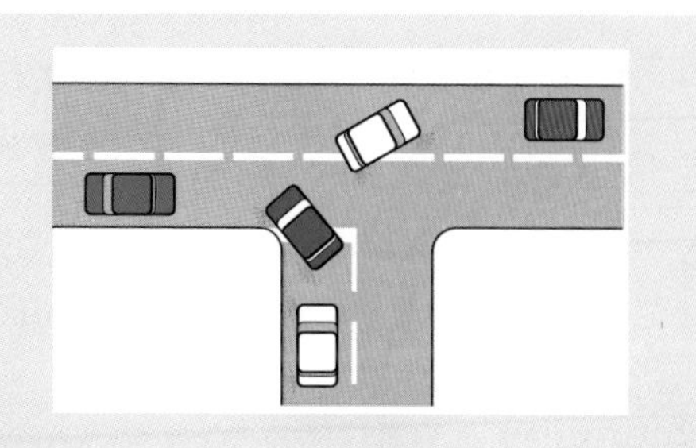

Their record started like this (R = right, L = left)

R R L R L L R R L R R R L R L R L L R R L R R L R

L L R R R L R L R R L R R R R L R R L L R R R R L ...

There is no pattern here. You cannot predict what the next turn will be. The outcome of each turn is **uncertain.**

But you may notice that R happens more often than L.

50 turns were recorded.

31 of them were right turns.

Although you cannot predict which way a car will turn, it is **more likely** to turn right than left at this junction.

The **relative frequency** of right turns was $\frac{31}{50}$.

This is an estimate of the probability of a right turn at this junction.

B1 Here is a record made at another road junction.

R L L R L L L R L L R L L R L L L L R L L R L L R

L L L R L L R L L R L L R L R L L R L L L R L R L ...

Estimate the probability of a right turn at this junction.

B2 Gavin sells ice cream in three flavours: vanilla, strawberry and chocolate.
He keeps a record of the flavours people ask for.
It goes like this.

S C S V V S C S V C S V S S C V V S C V

V S S V S C S C S V C S C V S S C S V S

Estimate the probability that a person will ask for:

(a) vanilla (b) strawberry (c) chocolate

Check that the three probabilities add up to 1.

B3 Nina has a coin which she suspects is unfair.

She throws it 50 times. Here are the results.

H H T H T T H H T H H H T H T H T H H H T H T H H

H T H H T H H T H H T T H H H H T T H T H H T H H

What is the relative frequency of (a) head (b) tail

Does the coin seem to be unfair?

Changing a fraction to a decimal

To change a fraction to a decimal, divide the numerator (top) by the denominator (bottom).

$$\frac{23}{35} = 23 \div 35 = \boxed{0.6571428} = 0.66 \text{ to 2 d.p.}$$

B4 Fred is doing a project on newspapers. He stands by a newspaper stall and notes down which paper each person buys.

Here is his record after half an hour.

Paper	Tally
Mirror	𝍸 𝍸 𝍸 𝍸 ||
Sun	𝍸 𝍸 𝍸 |||
Express	𝍸 𝍸 ||
Mail	𝍸 𝍸 𝍸 ||
Star	𝍸 𝍸 |
Telegraph	𝍸 𝍸 𝍸
Times	𝍸 𝍸 |
Guardian	𝍸 𝍸
Independent	𝍸 |||

(a) Work out the relative frequency for each paper.
Write them as decimals, to two decimal places. Do they add up to 1?

(b) Estimate the probability that the next paper asked for is the *Sun*.

B5 Dervinia stood beside a road and noted how many people were in each car as it passed.

Here is a table of her results so far.

Number of people in car	1	2	3	4	5
Number of cars	62	93	51	27	17

Estimate the probability that

(a) the next car to pass has one person in it

(b) the next car to pass has at least four people in it

C Estimating probabilities

C1 Put 5p on 'Start'.

		Win 5p		
	Lose		Lose	
Win 5p		Start		Win 5p

Roll a dice and move one square.

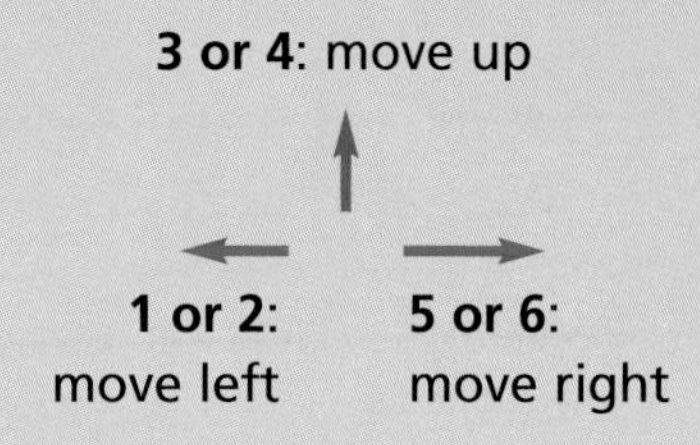

Then you roll the dice and move again.

If you land on 'Win' you get your own 5p back as well.

If you keep on playing, do you win or lose money?

Play the game many times and estimate the probability of winning.

C2 This is a game for two players.

One player (A) puts a counter on square A.

The other player (B) puts a counter on square B.

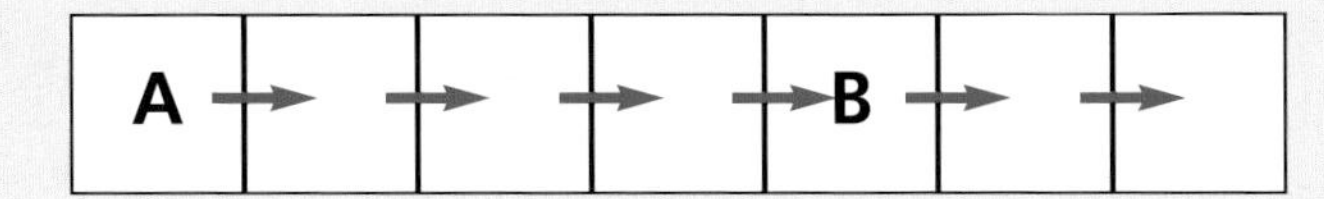

Throw a dice.

Odd: A moves 2 squares
Even: B moves 1 square

A wins if they catch up with B, before B gets to the last square.
(Or if they overtake B.)

Otherwise B wins.

(a) Which player do you think is more likely to win?

(b) Find by experiment an estimate of the probability that A wins.

(c) What if there were an extra square?

Project work

Design a simple dice-and-board game for two players.

Play the game a number of times and estimate the probability of each player winning.

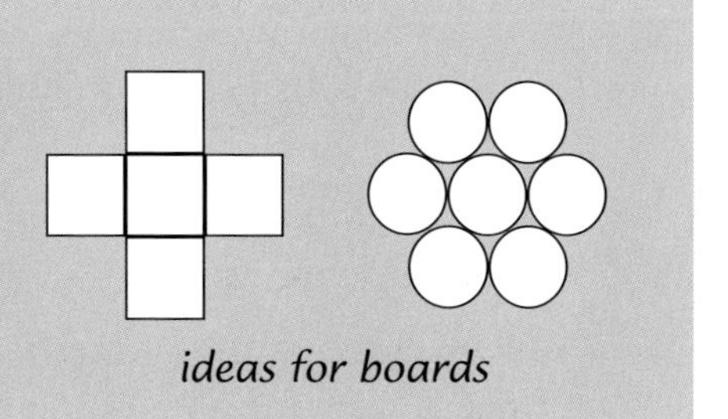

ideas for boards

D Simulation

Each box of these cornflakes contains a model. Suppose all three models are equally likely to turn up in a packet.

How many packets do you have to buy to get the set?

You may be lucky! You may get all three from the first three packets you buy.

But you could go on buying packets for ever and never get the set!

We can study this situation by using dice instead of cereal packets. This is called **simulating** the situation.

D1

1. Call the three models A, B and C.

   When you roll the dice, let

   1 or 2 represent A

   3 or 4 represent B

   5 or 6 represent C

2. Run the simulation until you have 'collected' all three models.

   Note down how many packets you 'bought'.

3. Re-run the simulation many times.

   Keep a tally of the number of packets 'bought'.

   Use your data to estimate the probability that you will need to buy 6 or more packets to collect the set.

Number bought	Tally
3	
4	
5	

D2

1. Some people would like to have both boys and girls in a family.

   Suppose the probability that a new-born baby will be a boy is $\frac{1}{2}$.

   (Actually it is slightly more than $\frac{1}{2}$.)

2. Use a dice to simulate the births of children in a family.

   You could use 'odd' for a boy and 'even' for a girl.

3. Throw the dice until there is at least one of each.

   For each run of your simulation, record how many children were born.

4. Estimate the probability that 4 or more births are needed before there is one of each sex.

E How often?

Relative frequency or probability can be used to estimate how often something will happen.

For example, the probability that a fair coin lands 'head' is $\frac{1}{2}$.
If the coin is thrown 50 times, we would expect about $\frac{1}{2}$ of 50, or 25 throws, to be heads.

E1 An ordinary fair dice is thrown 300 times.
About how many times would you expect to get (a) a six (b) an even number

E2 Jan did an experiment with a drawing pin.
In 100 throws, it landed point up 40 times.

(a) Estimate the probability that this drawing pin will land point up.

(b) If the same drawing pin is thrown 400 times, about how many times would you expect it to land point up?

E3 Karla dropped another drawing pin 100 times. It landed point up 44 times.

(a) Estimate the probability that this drawing pin will land point up.

(b) If Karla drops the pin 250 times, about how many times would you expect it to land point up?

E4 Steve made a solid shape whose faces were hexagons and pentagons. He rolled it 80 times. It stopped on a hexagonal face 56 times and on a pentagonal face 24 times.

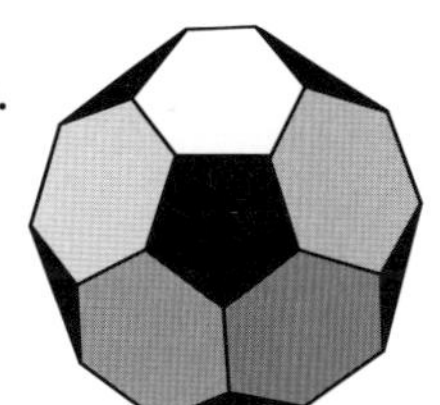

If the same solid is rolled 500 times, about how many times would you expect it to stop on a hexagonal face?

What progress have you made?

Statement	Evidence
I can estimate a probability using relative frequency.	1 A plastic square-based pyramid is thrown 50 times. It lands on its base 35 times. Estimate the probability that the pyramid lands on its base.
I can use probability to estimate how often something will happen.	2 A fair five-sided spinner is numbered from 1 to 5. If it is spun 400 times, about how often would you expect it to land on the number 4?
I can use simulation to estimate a probability.	The simulations you carried out in section D show this.

15 Bearings

This work will help you
- measure and record a direction as a three-figure bearing
- fix the position of a point by using its bearings from two other points

A Direction

T

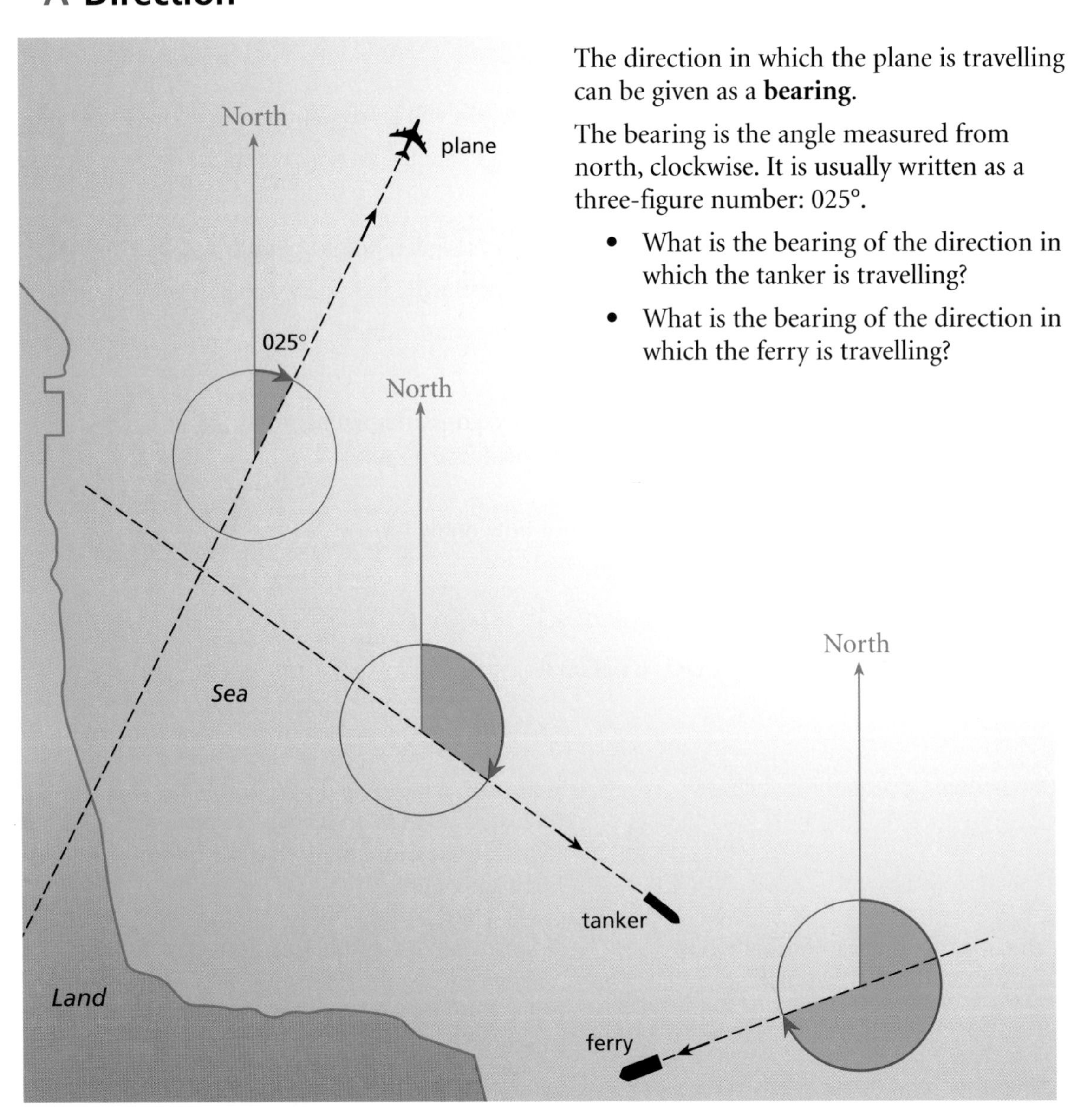

The direction in which the plane is travelling can be given as a **bearing**.

The bearing is the angle measured from north, clockwise. It is usually written as a three-figure number: 025°.

- What is the bearing of the direction in which the tanker is travelling?
- What is the bearing of the direction in which the ferry is travelling?

You need sheet 202.

A1 The dotted lines show the route of the ferry from Ballycastle to Church Quarter.
Measure the bearing of (a) the first part of the journey (b) the second part

A2 A boat leaves Ballycastle and travels on a bearing of 040°. Draw its path.

A3 A boat leaves Bull Point and travels on a bearing of 140°.
Draw its path and mark where it reaches the coast of the mainland.

A4 A boat sails from Cooraghy Bay on a bearing of 200°. Draw its path.

A5 A helicopter flies over Rue Point on a bearing of 310°.
Will the pilot see Bull Point on her left or her right?

A6 A speedboat leaves Kinbane and travels on a bearing of 075°.
Draw its path. (You will need to draw a line going north through Kinbane.)

A7 Draw the line from Colliery Bay on a bearing of 330°.
A company wants to build an oil platform 6 km from Colliery Bay along this line.
Mark the position where they want to build the platform.

A8 A buoy is 3.5 km from Bruce's Castle on a bearing of 175°. Mark its position.

A9 Measure the bearing of the line going from Killeany to the Coastguard Station.
We call this **the bearing of the Coastguard Station from Killeany**.

A10 Draw the line from Doon to Rue Point and measure the bearing of Rue Point from Doon.

A11 Find the bearing of

(a) Rue Point from Cooraghy Bay (b) Ruecallan from Church Quarter

A12 A boat leaves Ballycastle. It first sails on a bearing of 030° for 5 km.
It then sails for 2 km on a bearing of 110°. Draw its path.

Measure how far it is now from Ballycastle and its bearing from Ballycastle.

A13 A yacht is in distress. Coastguards record its bearing from Doon (080°)
and its bearing from Cooraghy Bay (140°).

Draw the line from Doon on a bearing of 080° and the line from Cooraghy Bay
on a bearing of 140°. Mark the position of the yacht.

A14 A boat is drifting out of control.
Coastguards record its bearings from Bull Point and from Rue Point.

Time	Bearing from Bull Point	Bearing from Rue Point
2 p.m.	170°	235°
3 p.m.	140°	270°

Mark the two positions of the boat. Mark the point on the coast of Rathlin Island
that the boat is heading towards.

B On the moors

You need sheet 203.

B1 The scale of the map on sheet 203 is 4 cm to 1 km.
If you want to draw a line on the map to represent 3 km, how long will you draw it?

B2 What length in centimetres represents

(a) 4.5 km (b) 2.8 km (c) 1.6 km (d) 4.9 km (e) 0.8 km

B3 A line on the map is 16 cm long. How many kilometres does it represent?

B4 What is the real length of a line which on the map is of length

(a) 8 cm (b) 15 cm (c) 7.2 cm (d) 10.4 cm (e) 5.4 cm

To avoid calculations of distances, you could make a 'scale ruler' by copying the scale of the map along a straight edge (for example, a folded piece of paper).

B5 Janice goes for a walk on the moor. She starts at Ridgeway Cross.
She walks for 2.8 km on a bearing of 064°.
Then she walks for 2.2 km on a bearing of 335°.

(a) Draw her walk. (b) How far is she from the nearest road?

B6 Grant starts at Portford Bridge. He walks for 1.6 km on a bearing of 208°.
Then he walks for 1.2 km on a bearing of 261°.

(a) Draw his walk. (b) What is the bearing of Withypool Cross from his final position?

B7 Karina starts at Anstey Gate. She walks in a straight line to White Post.

(a) What bearing does she walk on? (b) How far does she walk?

B8 Green Barrow is on a bearing of 130° from Landacre Gate and on a bearing of 010° from White Post. Mark its position on the map.

B9 There is a disused quarry on a bearing of 077° from Upper Willingford Bridge and 338° from Cloggs Farm. Mark it on the map.

B10 (a) Copy and complete this table of bearings.
(CF = Cloggs Farm, CG = Coombe Gate, OB = Old Barrow, PP = Porchester's Post, RC = Ridgway Cross, RF = Red Ford, W = Willingford, WC = Withypool Cross, WP = White Post)

RC to CF		CF to RC	
CG to OB		OB to CG	
PP to W		W to PP	
WC to WP		WP to WC	
RF to OB		OB to RF	

(b) Can you see a connection between 'the bearing from X to Y' and 'the bearing from Y to X'? Describe the connection.

*__B11__ Steve is on the moor. From where he is, Portford Bridge is on a bearing of 030° and Willingford is on a bearing of 320°. Find and mark Steve's position.

C Bearings jigsaw puzzle

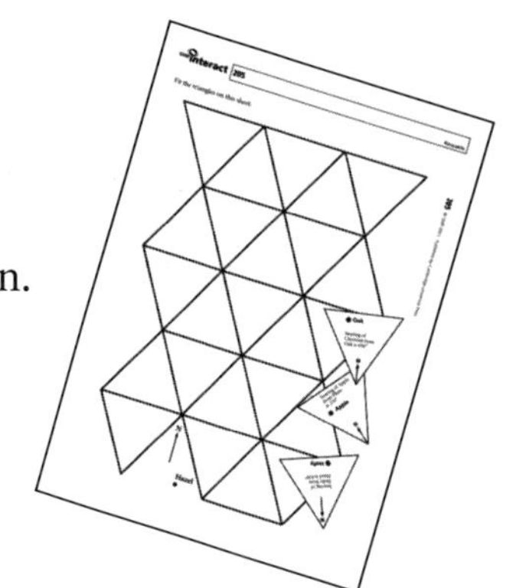

You need the jigsaw pieces from sheet 204, and sheet 205 to fit them on.

Each piece has the name of a tree on it.
Fit the pieces together following the information on them.

What progress have you made?

Statement	**Evidence**
I can measure and record a bearing.	**1** On the map below, measure (a) the bearing of Kirk Farm from Alcote (b) the bearing of Sterndale from Kirk Farm

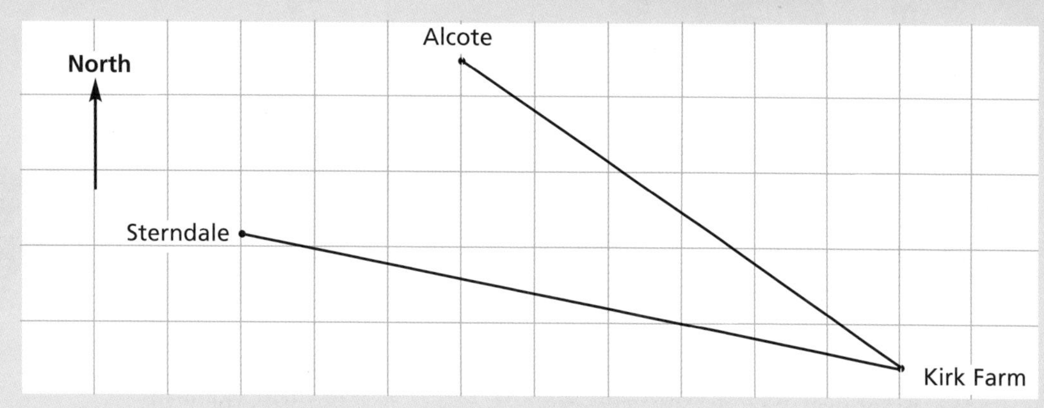

I can fix the position of a point given its bearings from two points.	**2** Draw the square ABCD below to a scale of 1 cm to 5 km.

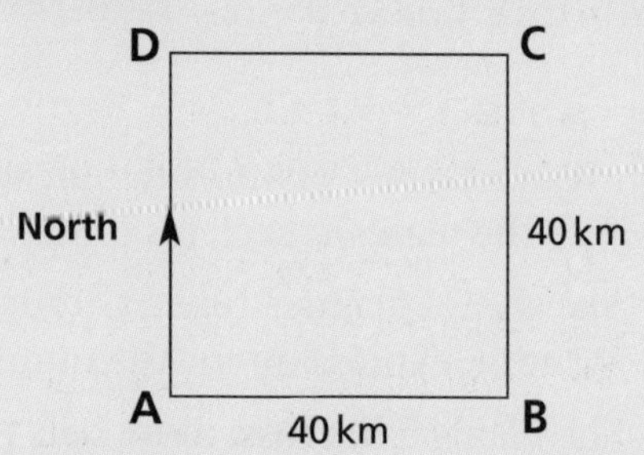

Find and label the point X whose bearing from A is 073° and from C is 202°.

I can plot a journey given the distance and bearing of each stage.	**3** Draw the following journey on the same diagram as for question 2. D to P: 33 km on a bearing of 118° P to Q: 28 km on a bearing of 232°

16 Forming equations

This work will help you

- translate statements in words into mathematical symbols
- form equations to help solve practical problems
- practise solving equations

A Expressions

Using algebra can help you solve problems. But first you need to translate statements in words into mathematical symbols. This section will help you practise this skill before going on to solve equations in a variety of situations.

A1 Each year, a farmer plants different sorts of crops.
Suppose she plants n hectares of barley.

(a) She always plants 3 times as much wheat as she does barley.
Write an expression for the number of hectares of wheat she plants.

(b) She plants 5 hectares more oats than barley.
How many hectares of oats does she plant?

(c) She plants half as much linseed as barley.
How much linseed does she plant?

A2 In Eckerley Zoo there are some elephants.
Suppose that there are n elephants.

(a) There are three times as many camels as elephants.
Look at the expressions in the box on the right.
Which is the correct expression for the number of camels?

(b) There are four times as many rhinoceroses as elephants.
Pick an expression for the rhinoceroses.

(c) There are three less tigers than there are elephants.
Which is the expression for the tigers?

(d) There are twice as many hippopotami as elephants.
There are three more giraffes than hippos.
Pick an expression for the number of giraffes.

$2n + 3$
$n - 3$
$3 - n$
$n + 3$
$3n$
$2n - 3$
$4n$
$n + 4$
$\frac{n}{4}$

A3 Eckerley Zoo has m lions.
Write sentences in words for each of the mathematical sentences below.

For example, for A you could write:

There are twice as many tigers as lions.

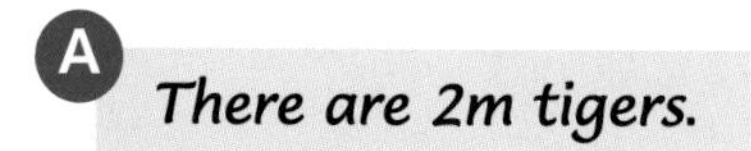

A
There are 2m tigers.

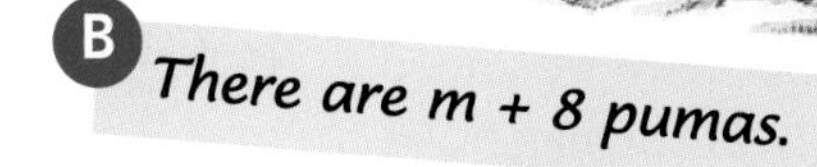

B
There are m + 8 pumas.

C There are $m - 2$ leopards.

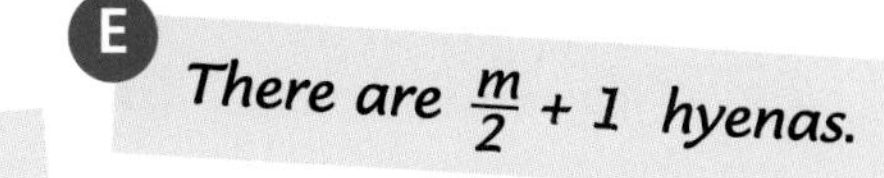

E
There are m/2 + 1 hyenas.

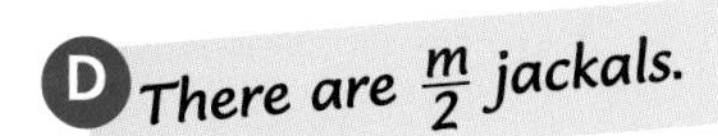

D
There are m/2 jackals.

A4 Suppose the zoo has p antelopes.
Invent four sentences of your own, using words, about how many animals there are in the zoo compared with the number of antelopes.

For each word sentence, write a mathematical sentence using letters.

A5 There is a queue of cars in a traffic jam.
Each car has four people in it.

Suppose there are c cars in the queue.

(a) How many people are in the cars altogether?

(b) Six cars turn round and leave the traffic jam.
How many people are there now?

A6 A minibus can carry nine people.
Suppose there are m minibuses.
They are all full, except one which has only four people in it.

How many people are in the minibuses altogether?

A7 A freight train is made up of one engine plus a number of trucks.
One truck, together with its load, weighs exactly 100 tonnes.
The trucks are each 12 metres long.
The engine weighs 80 tonnes and is 18 metres long.

Suppose a freight train has x trucks.

(a) Write an expression for the weight of all the trucks, in tonnes.

(b) Write an expression for the weight of the whole train.

(c) Write an expression for how long the whole train is, in metres.

(d) 10 trucks are taken off the train.
How many trucks are there on the train now?
How many tonnes does the whole train weigh now?

A8 There are k kittens in a litter.
The kittens weigh 850 grams altogether.

(a) Write an expression for the mean weight of one kitten.

(b) A kitten that weighs 50 g is taken away.
What is the mean weight of the ones that are left?

A9 There are n apples in a bag.

(a) Three times as many are added to the bag.
How many are in the bag now?

(b) The apples in the bag are now put in two equal piles.
Two apples are taken from one pile and put on to the other.
Write expressions for the number of apples in each pile.

A10 I bought a piece of ribbon that was n cm long.
I cut 20 cm off the ribbon. Then I cut the rest into 10 equal parts.

Write an expression for the length of each part.

A11 A piece of wire l cm long is bent into a square.
Write expressions for

(a) one side of the square (b) the area of the square

***A12** (a) A rectangle has one side 6 cm and the other a cm.
Write expressions for

(i) the perimeter of the rectangle
(ii) the area of the rectangle

a cm

6 cm

(b) Another rectangle has one side 8 cm and perimeter p cm.
Write expressions for

(i) the other side of the rectangle
(ii) the area of the rectangle

8 cm

perimeter p cm

B Equations

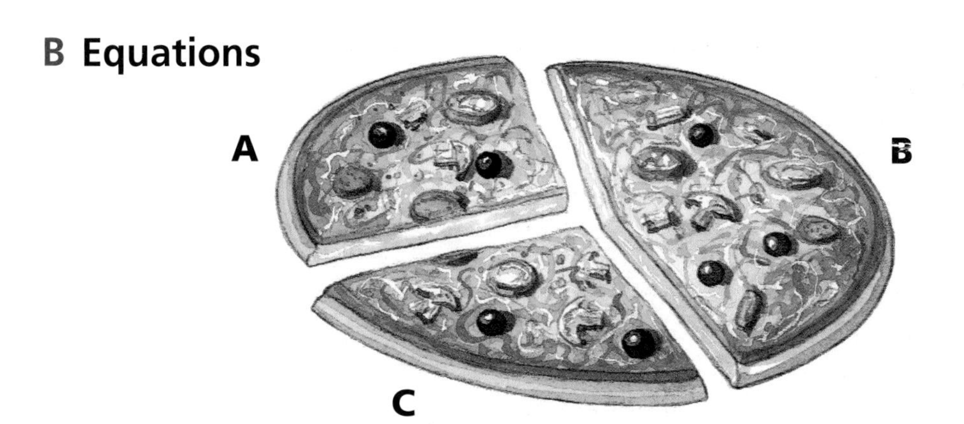

You can probably solve all the problems in this section without algebra. But the purpose of the section is to practise forming easy equations before you go on to forming and solving harder ones.

B1 Lava rock for rockeries costs £64 per tonne.

(a) Write down an expression for the cost of n tonnes of lava rock.

(b) Jonathan buys n tonnes of lava rock.
The rock costs him a total of £352.
Write this information as a mathematical equation.

(c) Solve your equation to find how many tonnes of rock Jonathan bought.
Check your answer works.

B2 Curtain material costs £7.60 per metre.

(a) What would the cost of x metres be?

(b) Jill buys x metres of curtain material. It costs her £57.

(i) Write the sentence 'x metres of material costs £57' as a mathematical equation.

(ii) Solve your equation to find how much material Jill bought.

B3 Serisa is packing discs into identical boxes.
She fills 8 boxes and has 13 discs left over.

Suppose each box holds d discs.

(a) Write an expression for the number of discs she started with.

(b) In fact, she started with 141 discs.

(i) Write an equation to show this.

(ii) Solve your equation to find the number of discs that fit into each box.

T

B4 Jon and Susan buy computer games.
Jon spends £12 more than Susan.

Suppose that Susan spends £s.

(a) Write down an expression for the amount that Jon spends.

(b) Write down an expression for the amount that the two of them spend altogether.

(c) The total bill is £73.
Form an equation, and solve it to find how much they each spend.

B5 Alan, Becky, Colin, Debbie, Eddie and Fran are comparing their weekly pocket money.

(a) Suppose Alan gets £x.
Here are some expressions for different amounts of pocket money.
Match each person's pocket money with the correct expression.

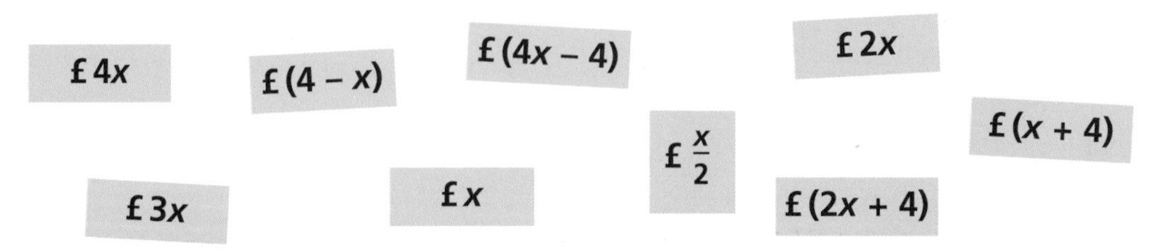

(b) Alan, Becky and Colin together get £13 in pocket money.
Form an equation, and solve it to find out how much Alan gets.

(c) Work out how much each of the six friends gets.

B6 I can buy single pieces of fruit from a stall on my way to work.

A pear costs 5p more than an apple.

A peach costs five times as much as an apple.

A banana costs 5p more than a pear.

A satsuma costs half as much as an apple.

An apricot costs 5p less than a peach.

(a) Suppose an apple costs c pence.
Look at these expressions for costs, in pence.
Which expression matches the cost of which fruit?
(You won't need all the expressions.)

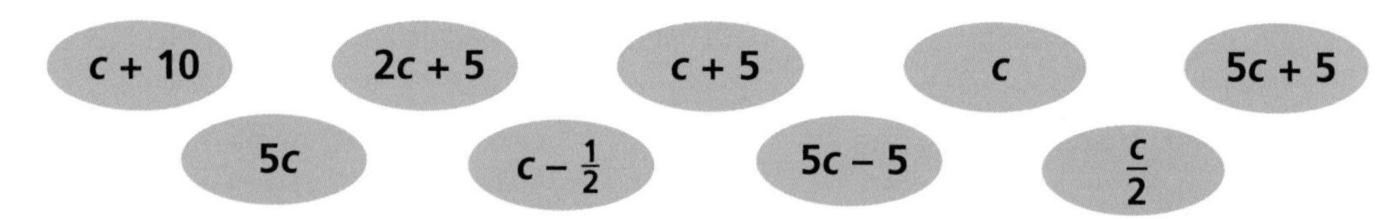

(b) A pear and an apricot together cost 60 pence.
Work out the cost of each of the six types of fruit.

B7 Coaches each have 53 passenger seats.

(a) A group of people occupy n coaches with 7 empty seats.
Which of the following expressions describes the number of people in the group? Explain your reasoning.

$n - 7$ $\quad$ $53 - 7n$ $\quad$ $53n - 7$ $\quad$ $53n + 7$

(b) Altogether, there are 470 passengers on the coaches.
Form an equation in n.
Solve it to find how many coaches are being used.

B8 Robin is four years older than Caroline.
Suppose Robin is x years old.

(a) Write down an expression for Caroline's age, in years.

(b) The sum of Robin's and Caroline's ages is 21.
Which of the following equations states this? Explain your choice.

$2x - 4 = 21$ $\quad$ $x + 4 = 21$ $\quad$ $2x + 4 = 21$ $\quad$ $x - 4 = 21$

(c) Solve the equation that is correct.
Work out Robin's and Caroline's ages.

Check your answers fit the original problem.

B9 The width of a rectangle is w cm.
Its length is double its width.

(a) Write down expressions for its length and its perimeter.

(b) The perimeter is 26 cm. Form an equation and solve it to find w.

B10 The width of a rectangle is x cm.
Its length is five times its width.

(a) Write down expressions for its length and area.

(b) The area of the rectangle is 245 cm^2.
Write down an equation and solve it to find x.

B11 A shirt costs £c. A sweater costs £13.11 more than a shirt.

(a) Write down an expression for the total cost of a shirt and a sweater.

(b) The price of a shirt and a sweater together is £43.21.
Form an equation, and solve it to find the cost of a shirt.

B12 A pair of boots costs £13.55 more than a pair of trainers.

(a) Choose your own letter to stand for the cost of a pair of boots, in £.
Write down an expression for the cost of a pair of trainers.

(b) The bill for a pair of boots and a pair of trainers was £98.93.
Form an equation and solve it to find the cost of a pair of boots.

B13 A mini-disc Walkman costs £29 more than a tape Walkman.
Suppose the cost of a tape Walkman is £m.

(a) Write down an expression for the cost of a mini-disc Walkman.

(b) A shop orders 22 tape Walkmans and 9 mini-disc Walkmans.
Write down expressions for

(i) the cost of the tape Walkmans

(ii) the cost of the mini-disc Walkmans

(c) The total cost of the order is £1594.
Form an equation and solve it to find the cost of each type of Walkman.

B14 In the cross shown, the length of AB is half the length of BC.

(a) If BC is l cm long, how long is AB?

(b) The total perimeter of the cross is 370 cm.
Form an equation in l and solve it to find the length of BC.

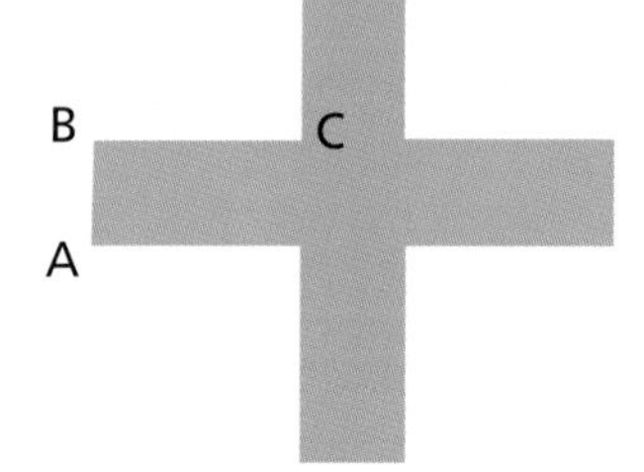

B15 Thomas is y years old.

(a) John is five years younger than Thomas; how old is John?

(b) Thomas is three years younger than Susan; how old is Susan?

(c) The combined age of the three of them is 40.
Form an equation and find the age of each person.

B16 I have n £20 notes and three times as many £5 notes in my wallet.

(a) How many £5 notes do I have?

(b) What is the value of all of the £20 notes?
What is the value of all the £5 notes?

(c) Write down an expression for the total amount of money in my wallet.

(d) Altogether I have £105 in my wallet.
Form an equation and solve it to find n.
How many of each type of note do I have in my wallet?

***B17** A money box contains some 5p coins and twice as many 10p coins.
Suppose there are f 5p coins.

(a) Write down an expression for the number of 10p coins in the box.

(b) Write down an expression for the total number of coins in the box.

(c) Write down a second expression for the total amount of money in the box.

(d) There are 33 coins altogether. Form an equation and solve it to find f.

(e) What is the total value of the coins in the money box?

***B18**

I think of a number.
I add 12. Then I add twice the original number.
Now I take off half of the original number.
My answer is 37.

Form an equation and solve it to find the number I first thought of.

***B19**

I think of a number.
I add to it one half of itself.
I then subtract one half of the sum.
My answer is 75.

Form an equation and solve it to find the answer to this puzzle.

C Harder problems

T

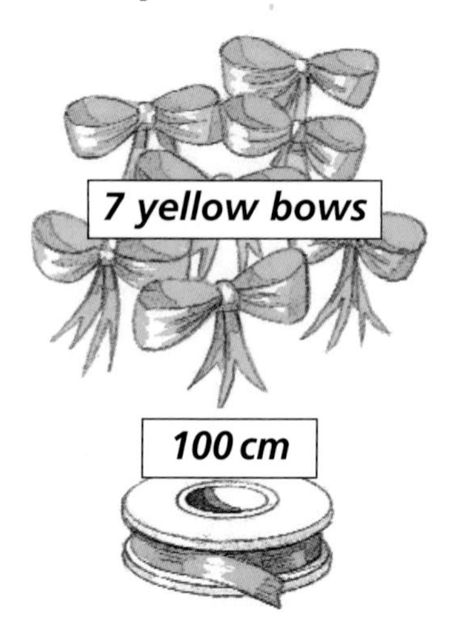

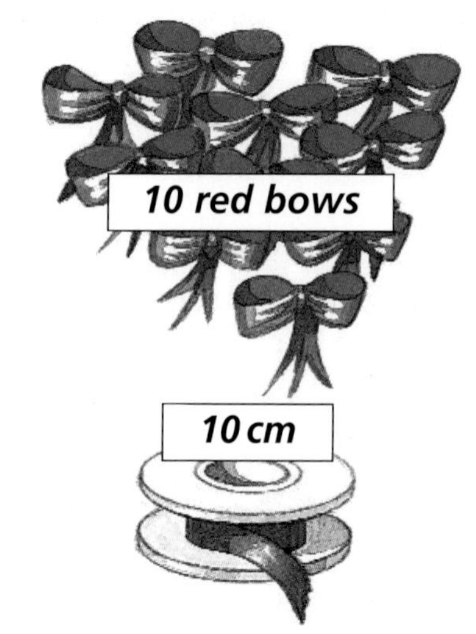

C1 Sarah and Juliet are both the same age (suppose they are x years old).

(a) Sarah multiplied her age by 3 and then added 27.
Write down an expression for the number she got.

(b) Juliet multiplied her age by 5 and then subtracted 4.
Write down an expression for the number Juliet got.

(c) They both ended up with the same answer. How old are they?

C2 Jim and Harry both think of the same number (call it n).
Jim multiplies the number by 3, and then adds 18.
Harry just adds 32 to the number.

(a) Write down expressions for the numbers they each ended up with.

(b) If they both end up with the same number,
what was the number they originally thought of?

C3 Zahir and Marisa were both thinking of the same number.
Zahir added 3 to the number and then multiplied the result by 8.
Marisa added 13 to the number and then multiplied the result by 6.
They were surprised to find that they both had the same answer.

What was the number they were both thinking of originally?

(Let the number they first thought of be x.)

C4 Alan and Emdad are both thinking of numbers.
Emdad's is 2 more than Alan's.
Alan multiplies his number by 2 and then adds 7.
Emdad subtracts 1 from his number and then multiplies the result by 5.

They both get the same answer. What were their original numbers?

C5 John is x years old.

(a) Ben is two years older than John. How old is Ben?

(b) John multiplies his age by 4 and subtracts 5.
Ben multiplies his age by 2 and adds 12.

They both get the same answer. How old is Ben?

C6 (a) Find the value of x that makes this triangle equilateral.

Check your answer by finding the lengths of all three sides.

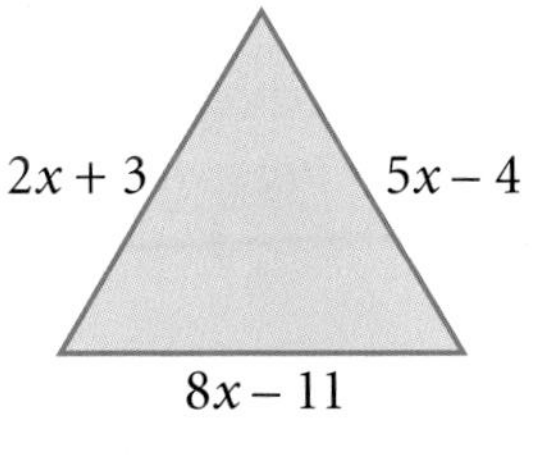

(b) Is it possible to find a value of x so that this triangle is equilateral?
If so, find the lengths of all the sides.
If not, explain why it is impossible.

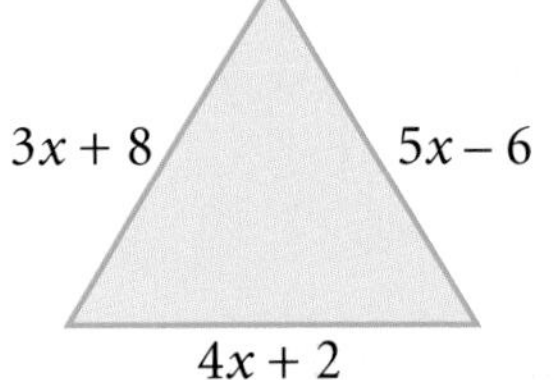

C7 This diagram shows the rates at which gas flows in a network of pipes
The rates are in cubic metres per minute.
The arrows show the direction in which the gas is flowing.

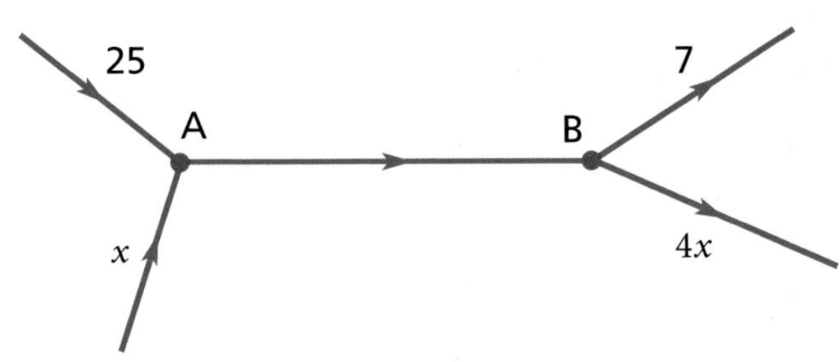

(a) Write down two expressions for the rate of flow of gas along AB.
(Assume that no gas is lost at the junctions.)

(b) Form an equation using x and solve it.

(c) At what rate is gas flowing along AB?

C8 This diagram shows another pipe network, similar to that in question C7.

At what rate is gas flowing along CD?

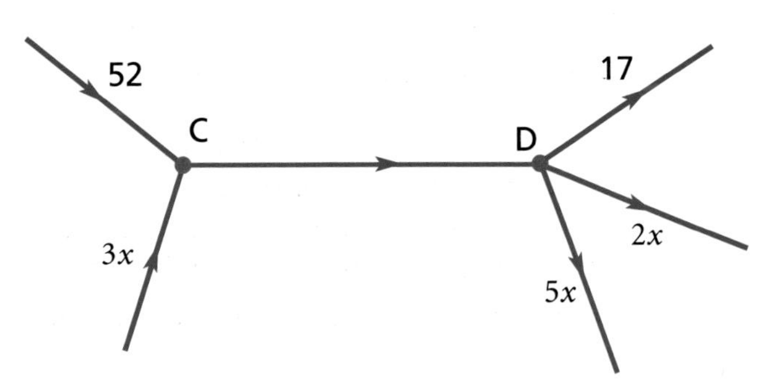

*C9 A carpenter is making shelves for a bookcase and a display cabinet. He has two equal planks from which to cut shelves.

From one plank he makes 5 shelves for the bookcase and has 20 cm left over.

From the other plank he cuts 3 shelves for the display cabinet and has 50 cm left over.

The display cabinet is 10 cm wider than the bookcase. How long was each plank?

What progress have you made?

Statement	Evidence
I can translate an English statement into mathematical symbols.	1 One kilogram of apples costs *a* pence. Pears cost 2 pence per kilogram more than apples. Write down an expression for the total cost of 3 kg of apples and 4 kg of pears.
I can form equations to help solve problems.	2 Today I spent *t* minutes doing my homework. Yesterday I spent 20 minutes less and tomorrow I expect to spend twice as long as today. I expect to spend a total of 2 hours 20 minutes over the three days. How long did I spend on my homework yesterday?
I can form and solve equations in which an unknown appears on both sides.	3 William and Brian are both thinking of the same number. William multiplies the number by 4 and then subtracts 3. Brian adds 5 to the number and then multiplies the result by 2. They both get the same answer. Form an equation and solve it to find the original number.

Review 2

1

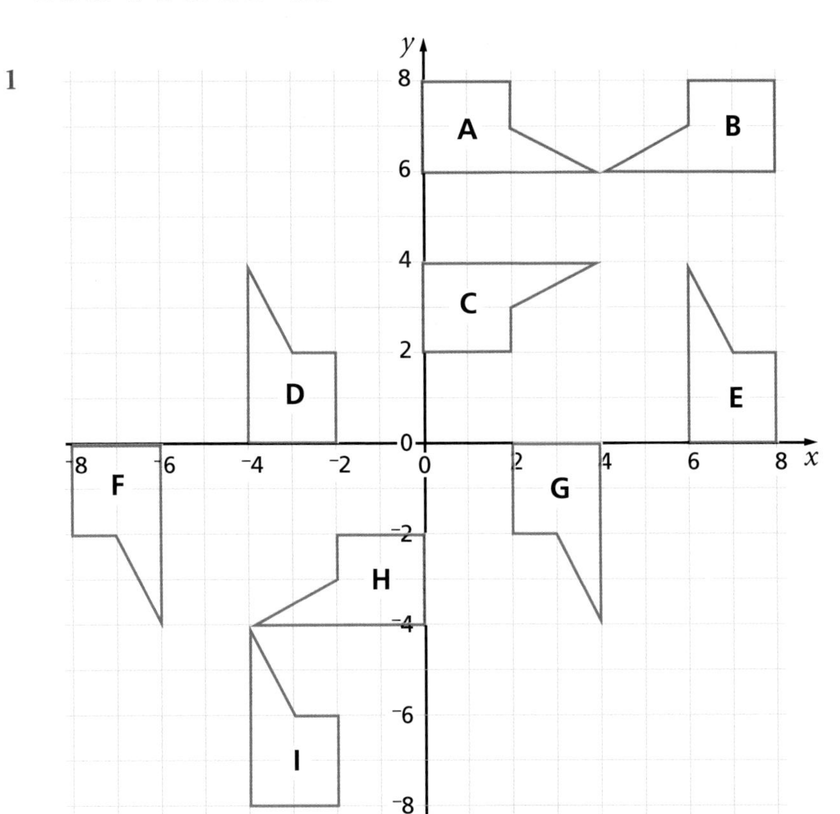

(a) Shape A can be mapped on to shape B by a reflection.
What is the equation of the mirror line?

(b) Shape D can be mapped on to shape C by a rotation.
Give the centre and angle of the rotation and say whether it is clockwise or anticlockwise.

(c) Describe fully the rotation that maps shape B on to shape C.

(d) What is the column vector of the translation that maps I on to E?

2 Use the same diagram as for question 1.
Describe fully the transformation that maps

(a) A to C (b) A to E (c) F to G (d) B to H
(e) G to C (f) I to F (g) A to F (h) C to E

3 (a) Copy and complete this statement.

To increase a quantity by 16%, multiply it by …

(b) A holiday firm increases the costs of its package holidays by 16%.
Calculate, to the nearest £, the new cost of a holiday that originally cost £488.

4 Find the equation of each of the lines a to g.

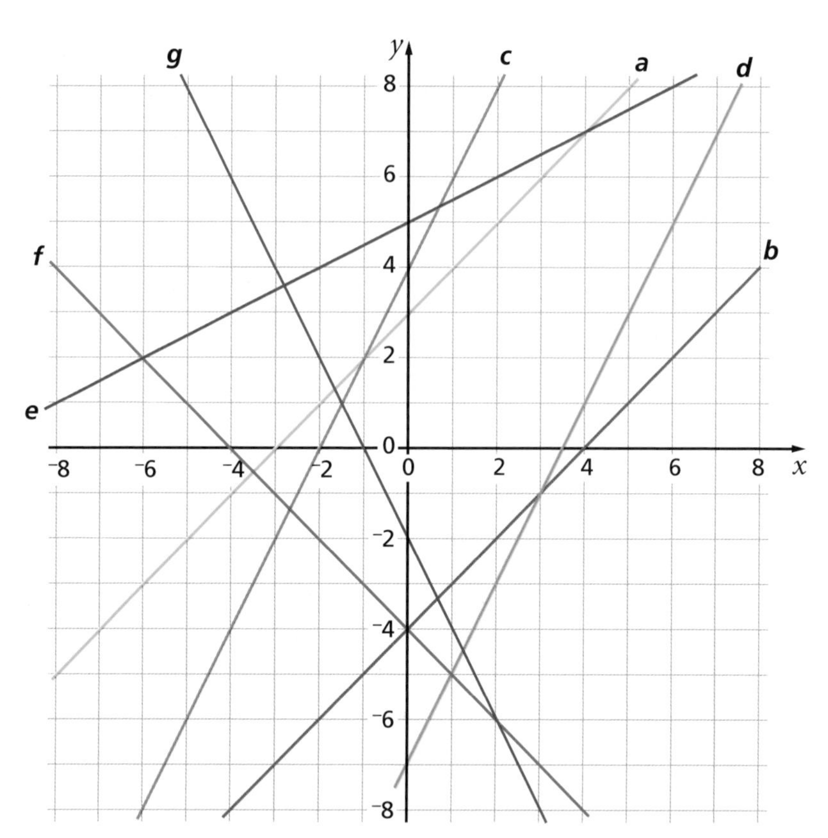

5 A multilink cube looks like this.

Sharmila rolled a multilink cube 80 times.
She recorded how many times it landed in each of these three positions:

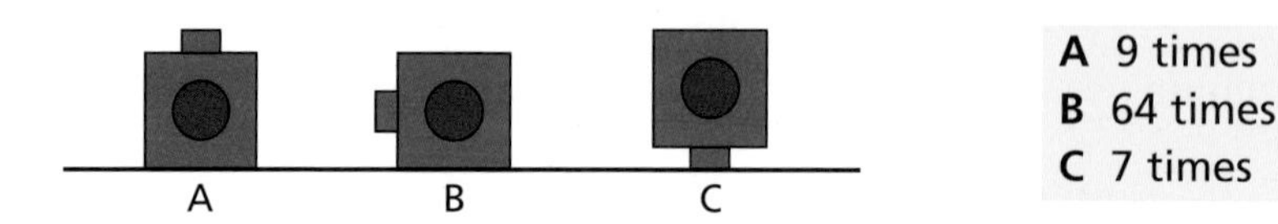

A 9 times
B 64 times
C 7 times

(a) Estimate, as a decimal, the probability that the multilink cube will land in

(i) position A (ii) position B (iii) position C

(b) Give a reason why position B is more likely than either of the others.

6 Simplify these expressions.

(a) $14 - 5x - 3 - 2x$ (b) $3x + 7 - 5x - 2$ (c) $x - 6 - 4x + 10$

(d) $4x - 3y + 2x - 1$ (e) $9 - 5x - 2y + 4 - y$ (f) $7x - 6y - 2 + 3y - 4x$

7 Draw axes on squared paper with x from 0 to 16 and y from 0 to 8.
Mark the points P(1, 1), Q(3, 0) and R(4, 2) and draw the triangle PQR.

Draw the enlargement of the triangle with centre (0, 0) and scale factor 3.
Write down the coordinates of the images of P, Q and R.

8 (a) Copy and complete this statement.

To reduce a quantity by 7%, multiply it by …

(b) A garage reduces the prices of secondhand cars by 7%.
Calculate, to the nearest £10, the new price of a car that cost £4320 before the reduction.

9 Lara has 20 rabbits. Some are white and the rest are brown.

Let n stand for the number of white rabbits Lara has.

(a) Write down an expression for the number of brown rabbits she has.

(b) Each white rabbit is worth £5 and each brown rabbit £3.
Write down an expression for the total value of all the rabbits in £.
Simplify your expression as far as possible.

(c) The total value of Lara's rabbits is £74.
Write down an equation involving n and solve it to find how many white rabbits Lara has.

10 A boat race has a triangular course ABC.
The bearing of each section is as follows:

from A to B 050°
from B to C 200°
from C to A 300°

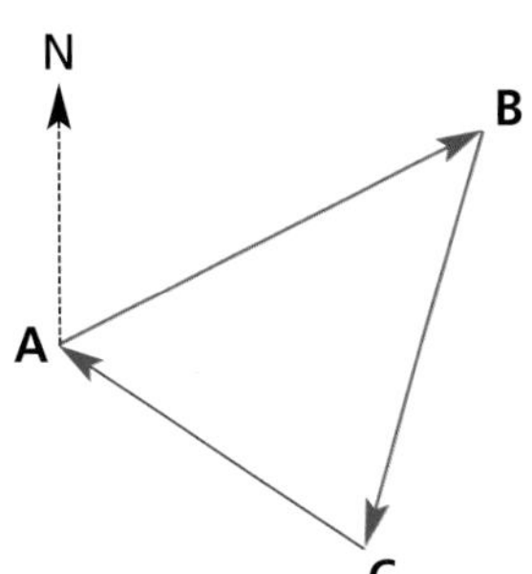

(a) Copy the sketch. Add north lines at B and C.
Mark and label the three angles given above.

(b) Work out the angles at the corners of triangle ABC (the angles ABC, BCA and CAB).

11 (a) Sethi's season ticket went up in January from £36.50 to £41.75.
What was the percentage increase, to the nearest 1%?

(b) When Jane changed jobs, the cost of her season ticket went down from £29.50 to £26.30.
What was the percentage decrease, to the nearest 1%?

12 Dawn had four times as many stickers as her brother Karl.
She gave eight of her stickers to Karl. Now she has twice as many as Karl.

Let n be the number of stickers Karl had to start with.

(a) Write down an expression in terms of n for

(i) the number Dawn had to start with (ii) the number Dawn has now

(iii) the number Karl has now

(b) Form an equation and find the value of n.

Ratio and proportion

This work will help you

- understand direct proportion and link it with graphs and equations
- do calculations involving ratio and proportion

A Some problems

T

1

Water comes out of a tap at a steady rate.
40 litres comes out in 3 minutes.
How much comes out in 15 minutes?

2

Pancakes
(makes 4 pancakes)

50 g plain flour
125 ml milk
15 g margarine
1 egg

How much of each ingredient will be needed for 12 pancakes?

3

Damson fool
(serves 4)

300 g ripe damsons
20 ml water
100 g sugar
20 g custard powder
150 ml milk

How much of each ingredient will be needed for 7 people?

4

Leek and potato soup
(serves 6)

40 g butter
500 g potatoes
750 ml vegetable stock
100 g cheese
3 leeks

How much of each ingredient will be needed for 15 people?

5

2 hectares is about the same as 5 acres.

About how many acres are there in 9 hectares?

6

Vivid orange
Mix red and yellow paint in the ratio 5 : 3.

How much red and how much yellow are needed to make 20 litres of vivid orange?

B Direct proportion: multipliers

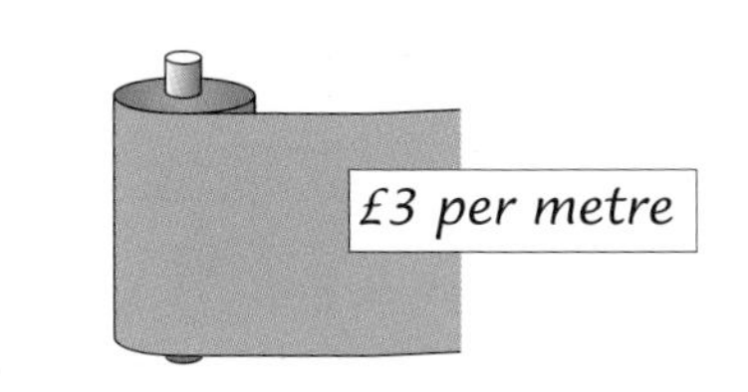

A shop sells material at £3 a metre.

The table below shows the cost of different lengths of material.
If you multiply the length by, say, 4, then the cost is multiplied by 4.

Length in metres	0	1	2	3	4	5	6	7	8
Cost in £	0	3	6	9	12	15	18	21	24

(× 4 from 2 to 8; × 4 from 6 to 24)

If I buy 5 times as much as you, then I pay 5 times as much, and so on.

We say that the cost is **directly proportional** to the length.

Here is another example of direct proportion.

- The weight of a pile of bricks is directly proportional to the number of bricks.

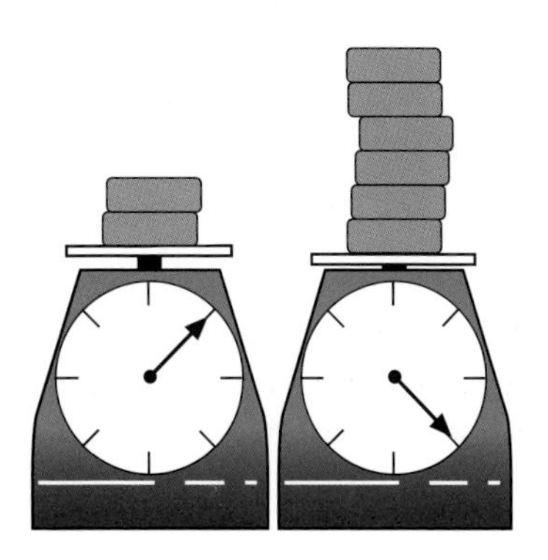

Do not use a calculator for these questions.

B1 The quantity of paint needed to paint a floor is proportional to the area of the floor.
Karl's floor has an area of 6 m^2. Jane's floor has an area of 30 m^2.
Karl's floor needed 1.5 litres of paint.

How much paint will Jane's floor need?

B2 The time taken to microwave a piece of meat is proportional to the weight of the meat.
Two pieces of meat weigh 250 g and 1 kg.
The smaller piece takes 20 seconds to microwave.

How long does the larger piece take?

B3 The quantity of water coming out of a tap is proportional to the time the tap is open.
In 20 seconds, 22 litres comes out of the tap.
How much comes out in 50 seconds?

B4 The length of a shadow is directly proportional to the height of the object (at a given time of day).

The shorter flagpole is 5 m tall, the longer 7.5 m tall.
The length of the shorter shadow is 9 m.

What is the length of the longer shadow?

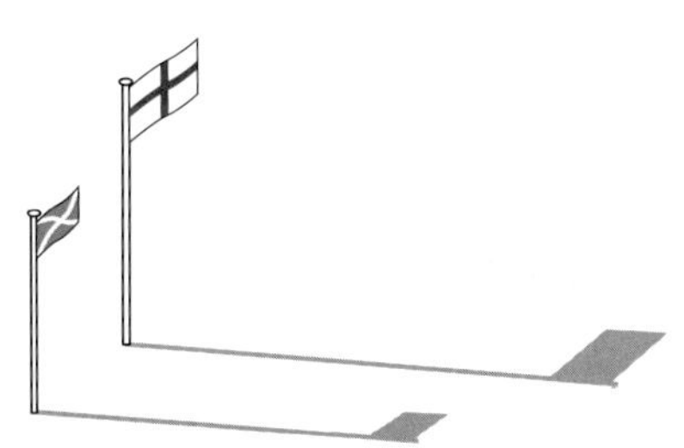

C Direct proportion: unitary method

Look at this problem.

5 identical bricks weigh 14 kg.
How much do 12 of the bricks weigh?

One way to solve the problem is to find the weight of **one** brick.

This method is called the **unitary method** ('unit' means 'one').

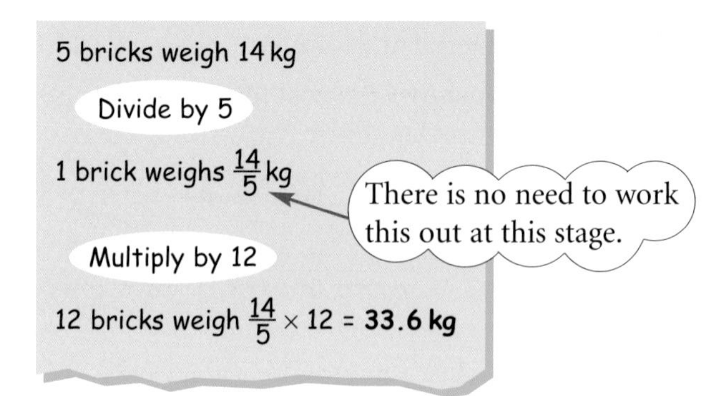

Do not use a calculator for questions C1 to C4.

C1 A photocopier prints 220 copies in 4 minutes.
How many copies will it print in 7 minutes?

C2 Emma wrote an essay of 2400 words. It took up 5 pages.
She has just written an essay that takes up 11 pages.
Estimate the number of words in this essay.

C3 Here is a recipe for liquid fertiliser.

Add 70 g of boric acid and 2 g of zinc sulphate to 5 litres of water.

How much boric acid should be added to 8 litres of water?

C4 These two pieces of metal are cut from the same sheet.
The smaller piece weighs 80 kg.
How much does the larger piece weigh?

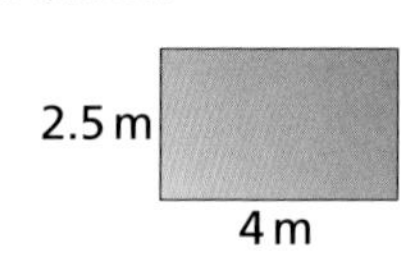

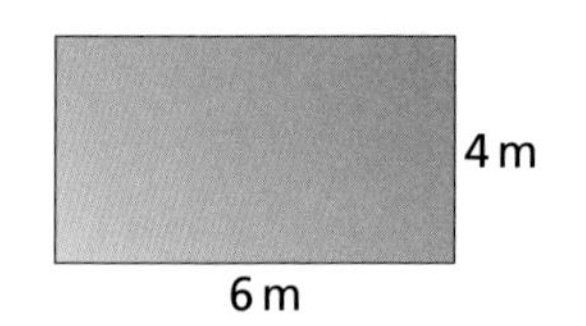

You may use a calculator for questions C5 to C8.

C5 A bookseller is packing copies of a textbook to send to schools.
24 copies of *Laugh-a-minute Maths* weigh 7.8 kg.

How much do 55 copies weigh, to the nearest 0.1 kg?

C6 Melanie used 6.5 kg of grass seed to plant a lawn of area 150 m^2.
She now has a 2.5 kg bag of grass seed. What area of lawn will she be able to plant?

C7 Josh painted a wall of area 35 m^2. He used 2.5 litres of paint.

(a) How much paint, to the nearest 0.1 litre, would Josh use to paint 50 m^2 of wall?

(b) What area of wall could he paint with 9 litres of paint?

C8 In March 2001, £1 was worth 3.30 New Zealand dollars.
How much was 20 New Zealand dollars worth in £ and pence (to the nearest penny)?

D Graphs and equations

The cost of material is directly proportional to the length.
The table below shows some lengths (L m) and costs (£C).

L (length in m)	0	2	4	6	8	10
C (cost in £)	0	5	10	15	20	25

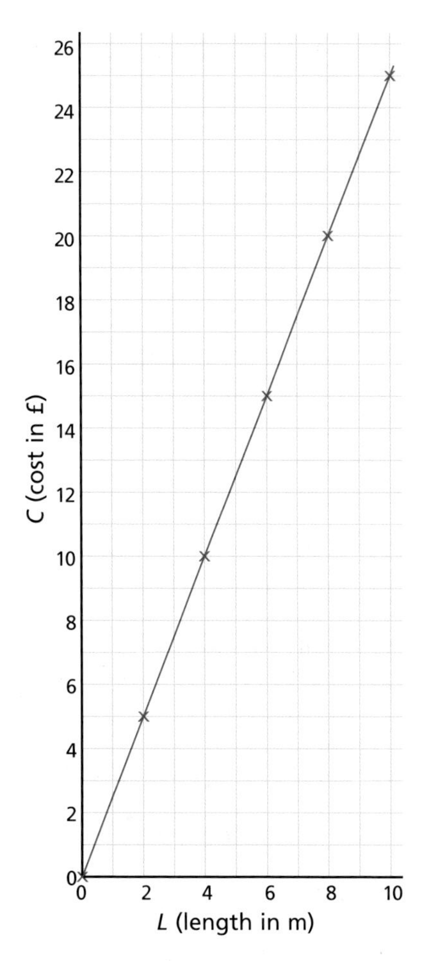

The graph of L and C is a straight line going through (0, 0).

The gradient of the graph is $\frac{25}{10} = 2.5$,

so the equation of the graph is

$$C = 2.5L$$

You can also see this from the table.
The cost in £ is 2.5 times the length in metres.

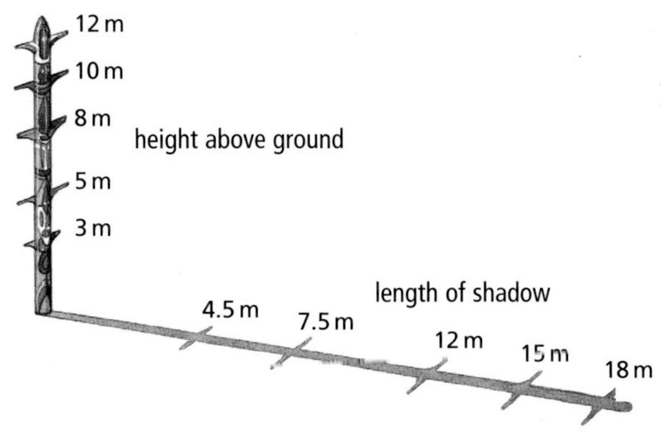

D1 This picture shows a totem pole and its shadow.

(a) Copy and complete this table of values of heights (h metres) and shadow lengths (s metres).

h	0	3	5	8	10	12
s	0	4.5				

(b) Draw a graph of h (across) and s (up).

(c) Find the equation connecting s and h.

D2 This table shows the number of copies produced by a printer when run for certain lengths of time.

t (time in minutes)	0	2	6	8	10
n (number of copies)	0	7	21	28	35

(a) What is the equation connecting n and t?

(b) Use the equation to find n when $t = 34$.

D3 This table shows the cost of a Southspeed train ticket for different journeys.

Distance (km)	0	15	19	28	40
Cost (£)	0	3.50	3.90	5.10	6.90

Plot the points from the table on graph paper and explain how you can tell that the cost is **not** directly proportional to the distance.

D4 Geri found the volumes ($V \text{cm}^3$) of some wooden blocks. She then weighed each block to find its mass (Mg).

Here are her results.

V	20	30	35	45	50
M	16	24	28	36	40

(a) Draw a graph of V and M.

(b) Find the equation connecting M and V.

(c) Use the equation to find

(i) the value of M when $V = 72$ (ii) the value of V when $M = 70$

E Conversion graphs

It is a well-known fact that 5 miles is approximately the same as 8 kilometres.

From this we can make a table of values …

miles	0	5	10	15	20	25
kilometres	0	8	16	24	32	40

… and draw a **conversion graph**.

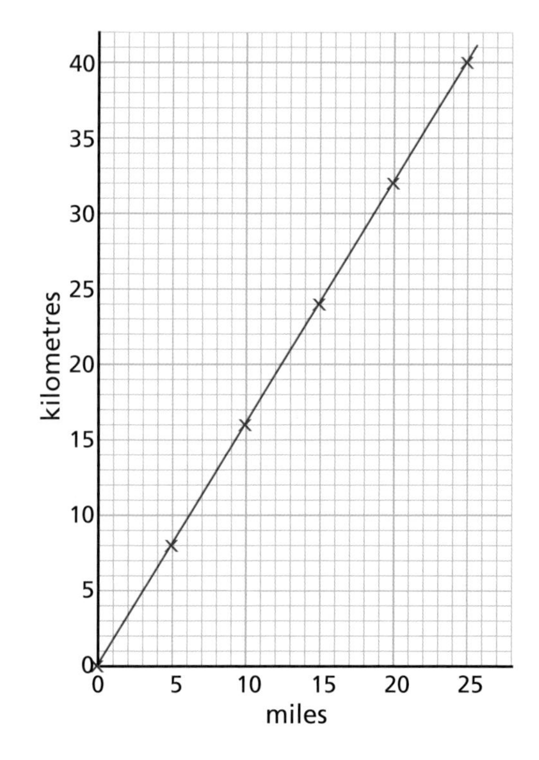

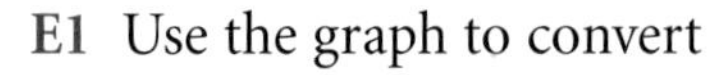

E1 Use the graph to convert

(a) 13 miles to km (b) 30 km to miles

(c) 24 miles to km (d) 22 km to miles

E2 The conversion graph only goes up to 25 miles.

(a) How can you still use it to help you convert 48 miles to kilometres?

(b) Use the graph to convert 100 km to miles.

E3 Ships' speeds can be measured in knots. 25 knots is approximately the same as 46 kilometres per hour.

(a) Using scales like those on the right, draw a conversion graph for knots and kilometres per hour.

(b) Use your graph to convert

(i) 5 knots to km per hour

(ii) 19.5 knots to km per hour

(iii) 20 km per hour to knots

(iv) 35.5 km per hour to knots

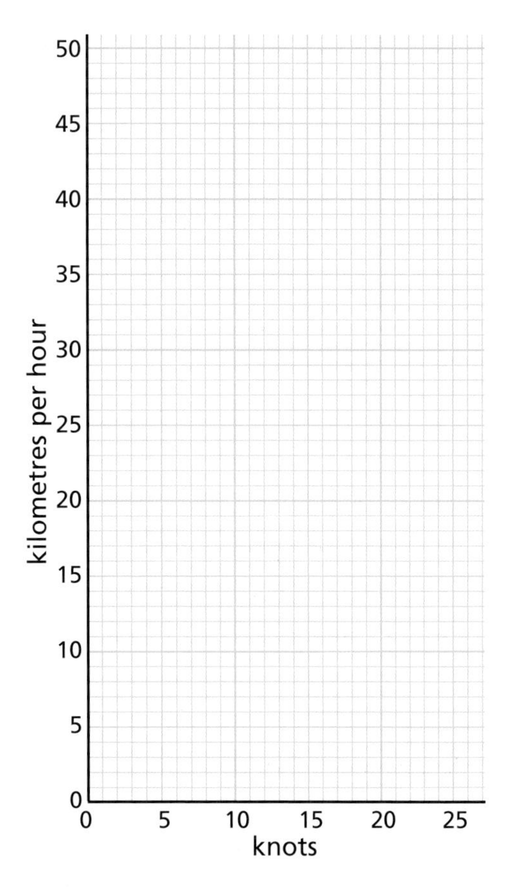

E4 Use the knots conversion graph to help answer these questions.

(a) A ferry travels at 13 knots. How far, in km, does it go in 4 hours?

(b) A launch travels at 22 knots. How far, in km, does it go in $2\frac{1}{2}$ hours?

(c) Convert the answer to (b) to miles.

E5 Large areas can be measured in square miles or square kilometres (km^2). 50 square miles is approximately equal to 130 km^2.

(a) Draw a conversion graph for converting areas up to 50 square miles.

(b) Use the graph to convert

(i) 37 square miles to km^2 (ii) 80 km^2 to square miles

E6 An airport shop shows its prices in pounds sterling (£) and in US dollars ($).

(a) From the information on this price ticket, draw a conversion graph for £ and $.

Zoom compact camera
£125 / $190

(b) Use the graph to find the missing prices on these tickets.

(i) Digital watch
£49 / $......

(ii) Make-up bag
£...... / $33

(iii) Perfume
£69 / $......

(c) According to the price tickets, what is £1 worth in US dollars?

F Ratios and fractions

F1 Donna has a collection of CDs.
The ratio of pop to classical CDs is 2 : 1.

What is the missing fraction in each statement below?

(a) The number of classical CDs is … of the number of pop CDs.

(b) The number of pop CDs is … of the total number of CDs.

(c) The number of classical CDs is … of the total number of CDs.

What is the missing number below?

(d) The number of pop CDs is … times the number of classical CDs.

F2 The ratio of girls to boys in a primary class is 3 : 2.

What is the missing fraction in each of these statements?

(a) The number of boys is … of the total number of children in the class.

(b) The number of boys is … of the number of girls.

(c) The number of girls is … of the total number of children.

What is the missing number below?

(d) The number of girls is … times the number of boys.

F3 In an adult drama club, $\frac{3}{4}$ of the members are women.

(a) What is the ratio of the number of men to the number of women?

(b) There are 15 women in the club. How many men are there?

F4 In a school chess club, $\frac{5}{8}$ of the members are boys.
There are 40 boys in the club. How many girls are there?

F5 A shop blends tea by mixing Ceylon and Kenya teas in the ratio 4 : 5.
What fraction of the mixture is Kenya tea?

F6 Last season, Brichester Rovers won 2 of their games, drew 3 and lost 1.

Find the following ratios.

(a) number of draws : number of losses

(b) number of wins : number of losses

(c) number of wins : number of draws

F7 Fifi and Gigi share an amount of money in the ratio 3 : 2.
Then Fifi gives half of her share to Gigi. What is the ratio of their shares afterwards?

F8 Green paint can be made by mixing blue and yellow.
Each 'box' on the opposite page describes a shade of green.

Put the boxes into groups so that the boxes in each group describe the same shade.
How many different shades are there?

A Mix blue to yellow in the ratio 2 : 3.

B

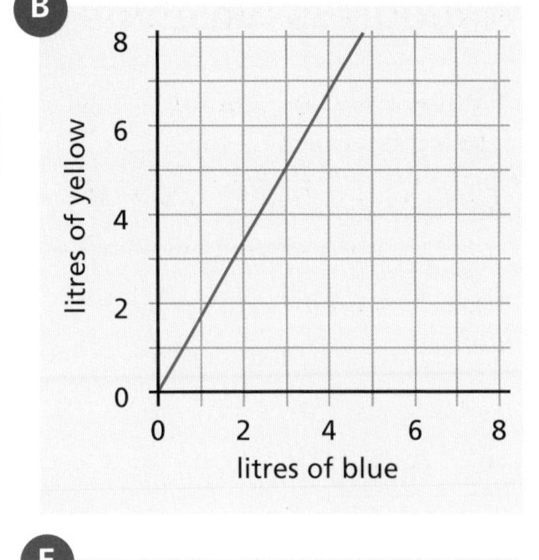

C $\frac{3}{8}$ of the paint you mix together is blue.

D

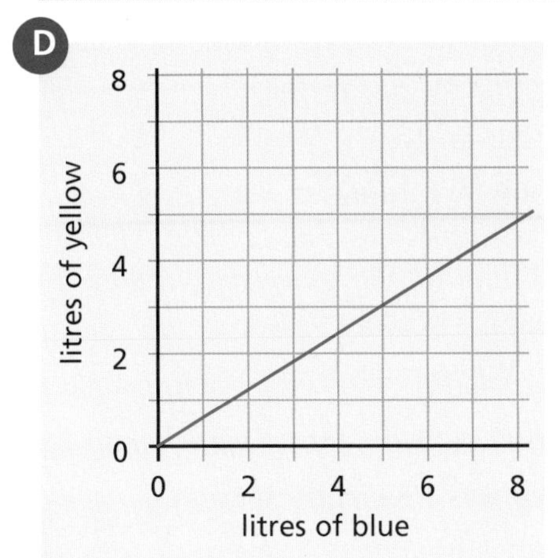

E

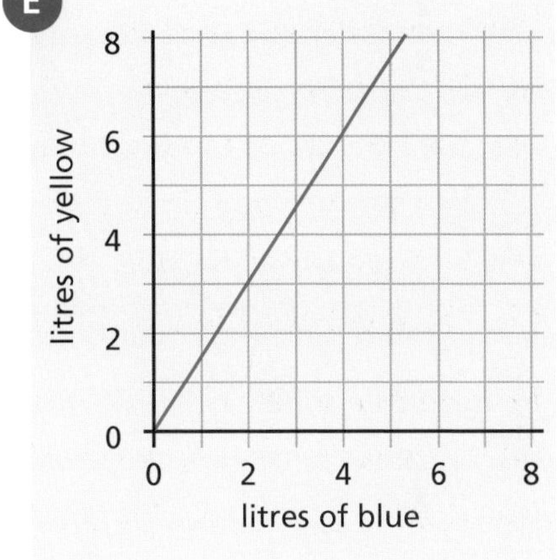

F

G

H

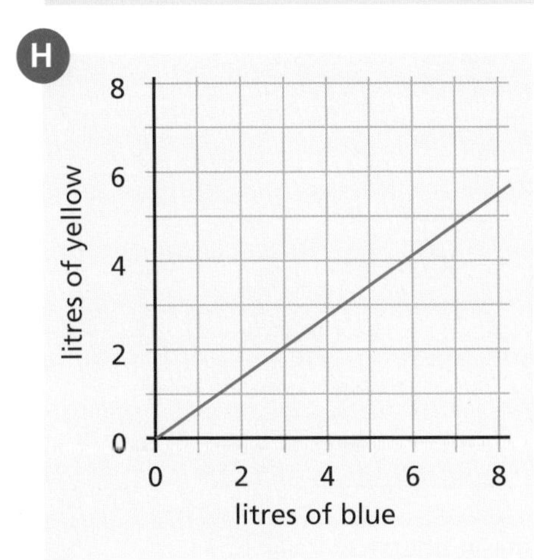

I $\frac{5}{8}$ of the paint you mix together is blue.

J Mix blue to yellow in the ratio 3 : 2.

K $\frac{3}{5}$ of the paint you mix together is yellow.

L If you mix b litres of blue and y litres of yellow, then $y = \frac{2}{3}b$.

M If you mix b litres of blue and y litres of yellow, then $b = 1\frac{1}{2}y$.

N If you mix b litres of blue and y litres of yellow, then $y = \frac{3}{5}b$.

O If you mix b litres of blue and y litres of yellow, then $b = \frac{2}{3}y$.

What progress have you made?

Statement	Evidence
I can use multipliers to solve problems involving direct proportion.	**1** The time taken to paint a floor is directly proportional to the area of the floor. Alan's floor has an area of $60\,m^2$. Bhuna's floor has an area of $90\,m^2$. Alan's floor took 50 minutes to paint. How long will Bhuna's floor take?
I can solve problems involving direct proportion using the unitary method.	**2** 13 bottles of wine weigh 15.6 kg. What do 30 of the same bottles weigh?
	3 In March 2001, 2.40 Swiss francs were worth £1. (a) How much was 1 Swiss franc worth, in pence (to the nearest 0.1 p)? (b) How much were 15.80 Swiss francs worth, to the nearest penny?
I can use a graph to find the equation connecting two quantities in direct proportion.	**4** The graph below shows the depth of water in a pool as it is being filled. What is the equation connecting D and T?

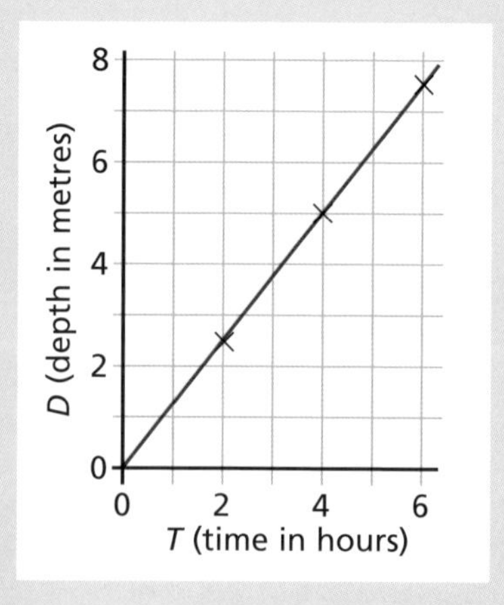

Statement	Evidence
I can link fractions and ratios.	**5** A piece of string is cut $\frac{5}{8}$ of the way along it. What is the ratio of the longer piece to the shorter piece?

18 No chance!

This is about probability.

The work will help you

- list outcomes systematically
- find probabilities in more complicated situations

A Beat the teacher

Coin game

I flip two coins.
If both land heads, you win.
If they are different, I win.
If they are both tails, we flip again!

Beat that!

I roll a dice. If I get a 6, it doesn't count, and I roll again until I get 1, 2, 3, 4 or 5.

Now you roll. If your score is bigger than mine, you win. If it's the same or less, then I win.

Dice difference

I roll two dice.
If the difference between the scores is less than 2, you win.
If it's 2 or more, I win.

- Is the teacher playing fair in each of these games?
  How do you decide?

B All the outcomes

B1 A game is played with these two sets of cards.
One card is picked at random from each set.

(a) Copy and complete this list of all the possible outcomes.

Heart	*Spade*
2	*2*
2	*3*

(b) What is the probability that

(i) both 2s are picked (ii) both cards have the same number

(iii) the spades number is greater than the hearts number

B2 A game is played with three coins.
The list of all the outcomes has been started here.

1st coin	*2nd coin*	*3rd coin*
H	*H*	*H*
H	*H*	*T*
H	*T*	*H*

(a) Copy and complete the list of outcomes. (You should have eight altogether.)

(b) What is the probability that the coins land as

(i) three tails (ii) two heads and one tail (in any order)

(iii) all three the same (iv) more heads than tails

(v) the same number of heads as tails

B3 A card is selected at random from a set containing the ace, king, queen and jack of clubs.
A counter which is black on one side and red on the other is also flipped.

(a) List all the outcomes for this.

(b) What is the probability that the card is a queen and the counter lands red?

(c) What is the probability of getting the counter black and either a king or queen?

B4 A game is played with three fair spinners.
Each spinner has an apple, a lemon and a bell on it.

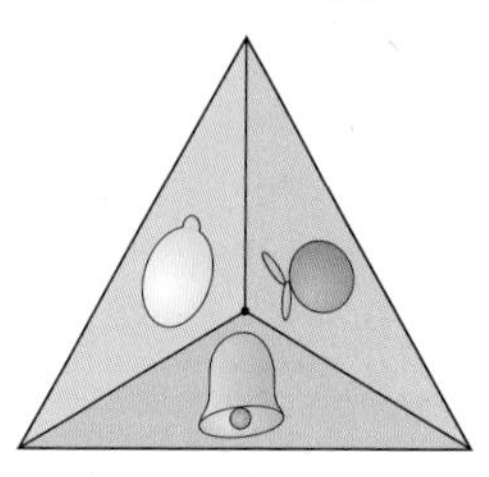
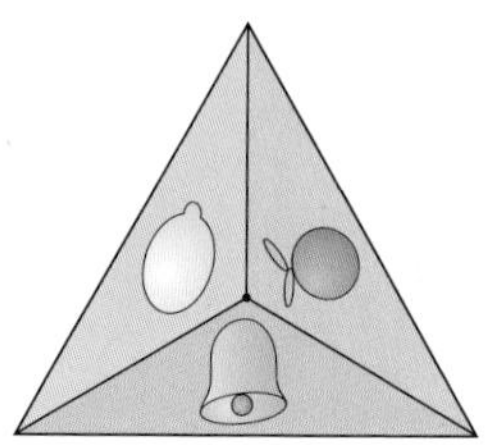
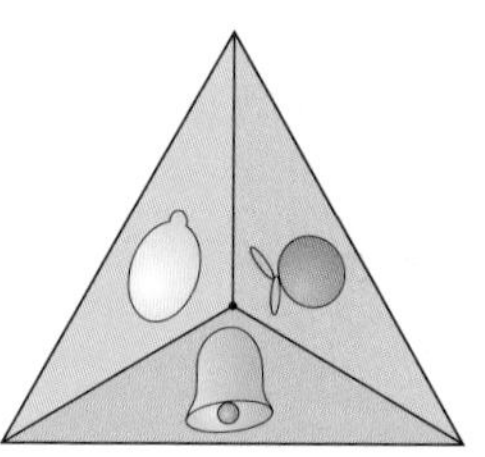

(a) List all the possibilities for the outcomes of how the spinners might land.

(b) Find the probabilities that the spinners land with

(i) three bells (ii) three the same

(iii) exactly two the same (iv) all three different

*B5 Some couples are considering the family they would like to have.
Assume that they are just as likely to have a boy as a girl.

(a) Tom and Anabelle say they want two children.
What is the probability they have (i) two girls (ii) children of different sex

(b) Sean and Mary think they want four children.
List all the possible outcomes under the headings 'First child', 'Second child' …

(c) What are the probabilities of Sean and Mary's children being
(i) all boys (ii) more girls than boys (iii) an equal number of each sex

*B6 When two coins are flipped there are 4 outcomes. With three coins there are 8.

(a) How many outcomes are there with (i) four coins (ii) five coins

(b) In Ireland the seventh son of a seventh son is said to have mystical powers.
How likely is a family of seven to be all boys?

(c) Look at the spinners in B4.
Can you find a rule for the number of outcomes for 3, 4, 5, … spinners?

C Easier ways of listing

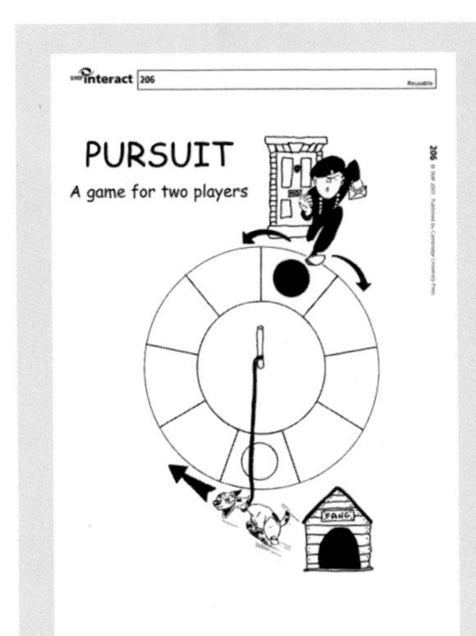

You need sheet 206, two four-sided dice and two counters of different colours.

- Postie is being chased by Fang.
Fang can only go clockwise.
Postie can go either way to escape.
- Fang catches Postie if either he lands on Postie or Postie lands on him.
- Postie goes first and rolls the two dice.
He can move, in either direction, the total score on the dice.
- Fang throws the dice and moves clockwise that score.
- Continue until Postie is caught.

In each of these, it is Postie's go.
(Postie is the red counter.)

Which way should he go, clockwise or anticlockwise?

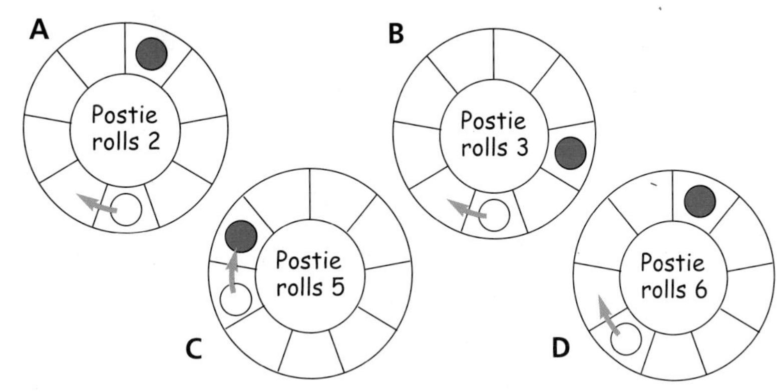

The outcomes of rolling two four-sided dice are: 1, 1   1, 2   1, 3   1, 4
2, 1   2, 2   … and so on.

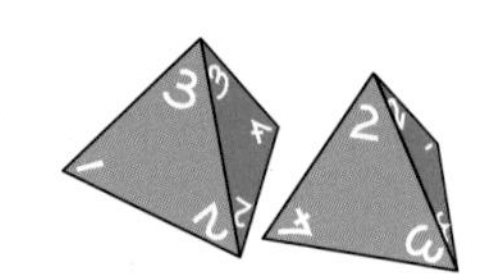

These outcomes can be shown as points on a coordinate grid.

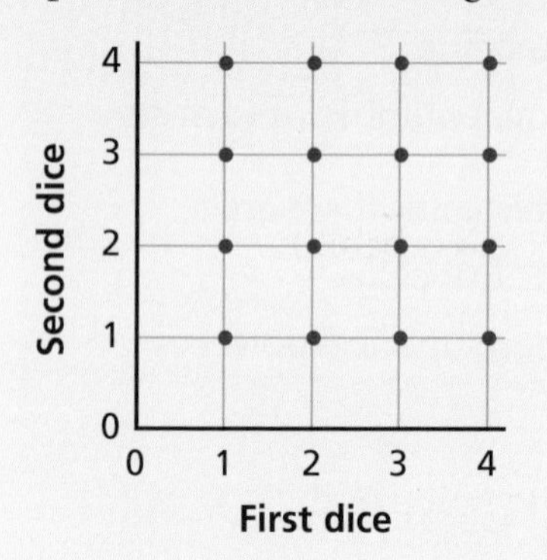

Or we can use a grid like this.

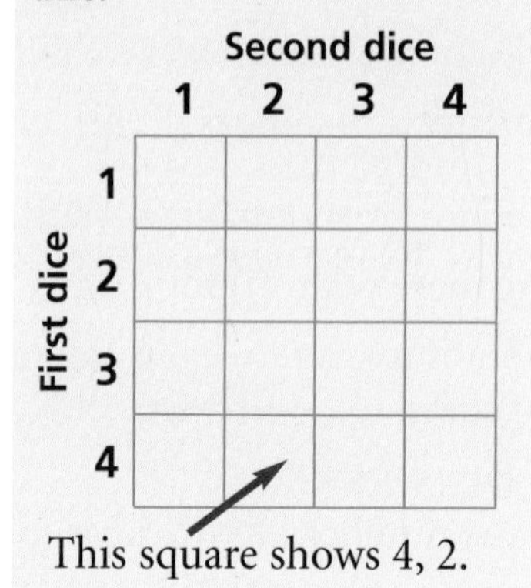

This square shows 4, 2.

The grid can be used to show, for example, total scores.

First dice \ Second dice	1	2	3	4
1	2	3	4	5
2	3	4	5	6
3	4	5	6	7
4	5	6	7	8

C1 Two four-sided dice are rolled.
From the last grid above, find the probability that the total of the two scores is

(a) 0   (b) 3   (c) greater than 3   (d) less than 3

C2 Copy and complete this grid to show all the total scores with two ordinary six-sided dice.

First dice \ Second dice	1	2	3	
1	2	3	4	
2	3	4	5	
3	4			

Find the probability that the total score is

(a) 7   (b) 10
(c) 2   (d) an odd number
(e) a prime number   (f) greater than 10

C3 (a) What score are you most likely to get with two six-sided dice?
(b) What scores are you least likely to get?
(c) Make a list of all the total scores and their probabilities.

C4 In another game, two six-sided dice are used, but instead of the total, the **difference** between the numbers is used.
For example, 3, 5 gives a difference of 2.

(a) Make a grid showing the differences for all the possible outcomes.
(b) Find the probability that the difference is

(i) 2   (ii) 1   (iii) 0   (iv) 4 or more   (v) 6

C5 This set of cards is used in a game called 'Salty Dog'.

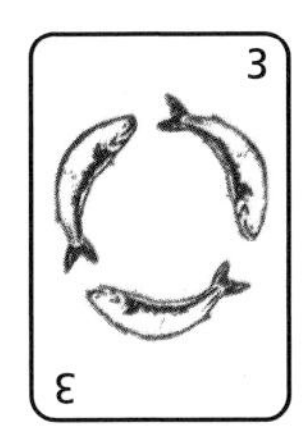

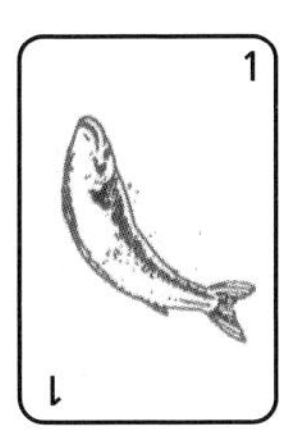

A blue set and a green set of the five cards are used.
One card is taken from each set.

Make a grid of all the possible outcomes and use it to find the probability that the two cards chosen

(a) have the same number
(b) are both multiples of 5
(c) are both less than 5
(d) have a total of 8 or more
(e) are both mermaids
(f) have 3 or less on the blue card

D Which is better?

Two classes have each designed a Wheel of Fate for a school fete.

They each charge 10p a go and have the same prizes.

Which would you rather have a go on?

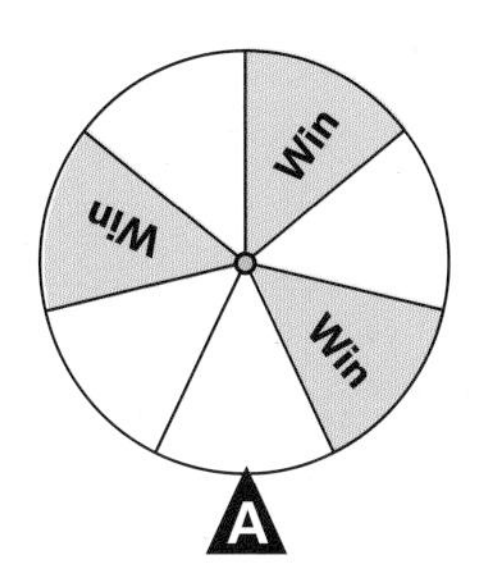

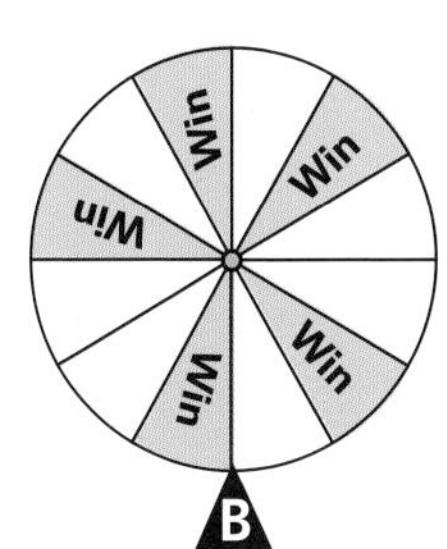

The probability of a win on A is $\frac{3}{7}$. On wheel B it is $\frac{5}{12}$.

Turning fractions into decimals, $\frac{3}{7} = 3 \div 7 = 0.428\ldots$ and $\frac{5}{12} = 0.416\ldots$

So you have a slightly higher chance of winning on wheel A.

D1 This is a scratch card.
You pick **one** box and scratch the surface.
You win if you find a star.

Be a Star!
Scratch just one box
Find a star and win £1

Below are three possible card designs with all the stars showing.
If you couldn't see the stars, which has the best chance of winning?

A
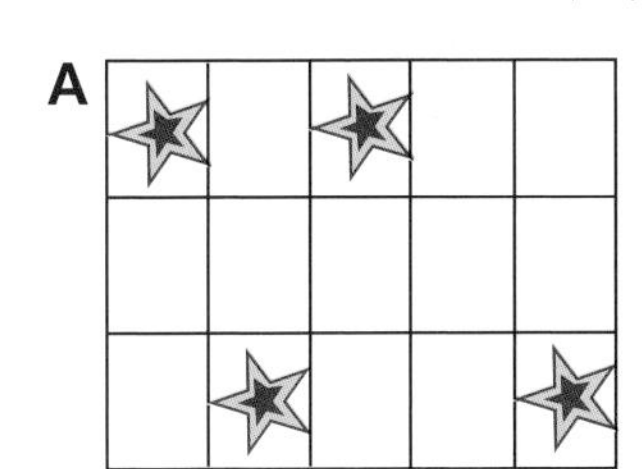
B
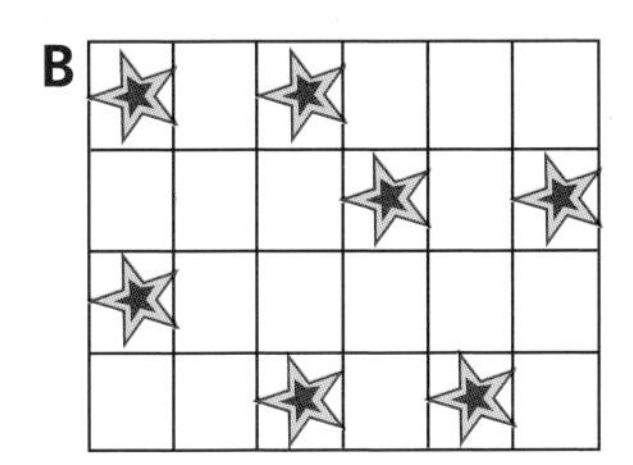
C
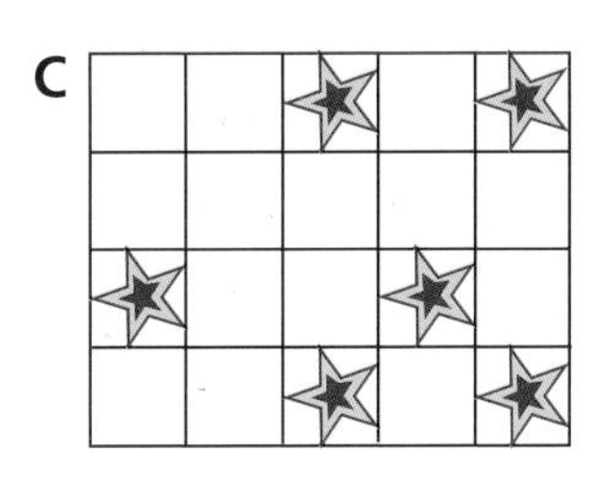

D2 Which of these is more likely? Explain why.

A Rolling a number greater than 4 on an ordinary dice

B Picking a weekend day when picking a day at random from the days of the week

D3 Which of these is more likely? Explain why.

A Getting more heads than tails when four coins are flipped

B Picking a heart from a pack of cards with no ace of clubs

D4 Find the probability of each of these events and say which is more likely.

A Getting a total score of 6 or more with two 6-sided dice

B Getting a total score of 12 or more with two 12-sided dice

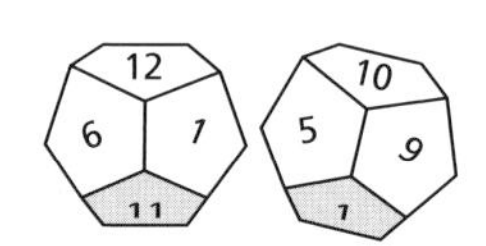

E How large, how many ...?

Working out how large things are can be difficult.
The two methods here are used by scientists to estimate large values.

Monte Carlo method

This method is used to estimate the area of awkward shapes.
It could be used to estimate the area of clover on a field.

Work with a partner.

- One person draws a 12 by 12 grid on squared paper. They draw a clover patch on the grid. This can be one patch or several smaller ones. They don't show it to the other person.
- The second person picks grid squares at random. They could use a 12-sided dice to generate pairs of coordinates. Or they could use the random number key on a calculator. They pick 20 squares.
- The first person says how many of these 20 squares are mostly clover.

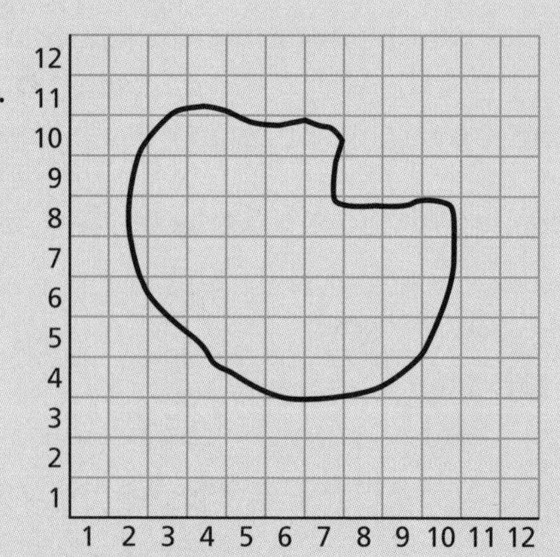

If, say, 14 out of the 20 squares were mostly clover, then roughly $\frac{14}{20}$ = 70% of the grid is clover.

Suppose each square represents 10 m by 10 m, so they each have an area of 100 m^2.
The total area of the grid is 14 400 m^2. So 70% of this is 10 080 m^2.

- Estimate the area of your clover patch in this way.
  Then compare it with the estimate you get by counting all the squares in the patch.

Capture–recapture method

Biologists often use this method to estimate the number of creatures in an area.

They go out and catch 20 creatures, say, and tag them in some way.

The population now looks like this. ➡

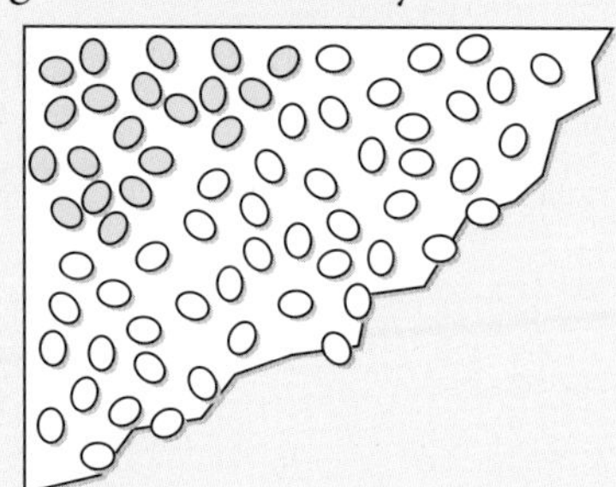

It has an unknown number of untagged creatures (O), and 20 tagged (O).

The creatures are allowed to mix together for a time.
Then another sample, say 50, are captured.

Suppose the second sample looks like this.

OOOOOOOOOOOOOOOOOOOOOOOOOOOOOOOOOOOOOOOOOOOOOOOOOO

In the second sample $\frac{4}{50}$ of the population are tagged.

If this is typical of the whole population, then

$\frac{4}{50}$ of the population $= 20$

and $\frac{1}{50}$ of the population $= 20 \div 4 = 5$

so the whole population $= 50 \times 5 =$ **250**

E1 An ornithologist wants to estimate the number of Canada Geese on a large field. She uses a net to catch some and catches 32 birds which she rings and sets free. The following week she manages to net 40 birds of which 8 are ringed.

(a) What fraction of her second catch are ringed?

(b) Find an estimate of the total number of birds on the field.

Try for yourself

Get a packet of dried peas or similar and place these in a large container.

Take a sample of 20 peas and mark them clearly.
Shake the container to mix in the peas thoroughly.
Without looking, take another sample of 50 peas and count the number of marked peas.
Use these figures to estimate the number of peas in the jar.

A cautionary tale ...

A biologist once tried to use this method to estimate how many crabs there were along a particular piece of shoreline. He captured some crabs and marked their backs with brightly coloured paint.

On returning a week later he found the beach littered with empty crab shells marked with paint. His paint had destroyed the crabs' camouflage and made them easy targets for gulls, who ate them!

What progress have you made?

Statement

Evidence

I can list all the outcomes in a situation and use these to find probabilities.

1 A spinner is divided into three sections with pictures of a star, a moon and a sun.

(a) List all the outcomes if three of these spinners are used.

(b) What is the probability that they show

(i) three stars

(ii) all the same

(iii) three different symbols

(iv) two or more with the same symbol

I can use a grid to show all the equally likely outcomes in a situation.

2 Two fair spinners are used with the numbers 1 to 5 on them.

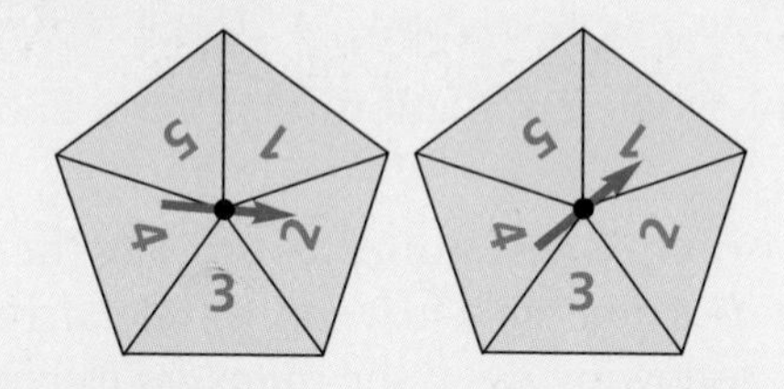

(a) Draw a grid to show all the equally likely outcomes.

(b) Find the probability that

(i) the spinners show the same number

(ii) the total of the numbers is 7

I can compare probabilities.

3 Which of the events below is more likely? Show how you decide.

A Getting a total of 5 or more with two four-sided dice

B Getting a total of 6 or more with two six-sided dice

19 Strips

This is about using algebra.
The work will help you

- simplify expressions
- use algebra to solve problems

A Review

A1 Find matching pairs of equivalent expressions.

A $6p + 3 - 5p + 2$

B $10 - 3p - 1 + 2p$

C $6p + 8p + 6 - p - 1$

D $7 + 3p - 4p - 2 + 2p$

E $11 - 9p + 4p - 2$

F $6 - 3p + 3 - 2p$

G $4 + 10p - 3p + 6p + 1$

H $7 + 6p + 2 - 7p$

A2 Simplify the following expressions.

(a) $9v + 4 - 5v - 3$
(b) $1 - 5w + 2 + 8w$
(c) $2x - 5x + 6x$
(d) $4 + 5y + 7 - y - 4y$
(e) $3 + 2k - 6k$
(f) $2 + 3j - 7 + 2j$
(g) $7 + 6h - 7h - 3$
(h) $5 + g - 2g + 4g$
(i) $4 - 8f - 1 + 2f$
(j) $5 - 4e - 3e + 3$
(k) $12d + 4 - 5d - 7$
(l) $2c - 5 - 9c + 10$
(m) $3 + 3b - 8b + 2b - b$
(n) $1 - 4a + 7 + 6a - 8$

A3 Simplify the following expressions.

(a) $2z + 4y + z + 2y$
(b) $4x + 7w + 3x - 4w$
(c) $6 + 3u + 6v - u + 1$
(d) $8s + t - 9s + 5t$
(e) $10 + 4q - 8r + 5q - 7$
(f) $5n + 12p - 7n - 9p + 3n$
(g) $10l + 3 - 7m - 5l$
(h) $11 - 4k - 5j - 3k - j$
(i) $7g + 4h - 9g - h$
(j) $3e + f - 5 + e - 7f$
(k) $3 + c - 4 + 6c - 5d$
(l) $7a - 5b + 2a - 9b + b$

B Addition strips

This is how to make an addition strip.

- Start with a strip of squares and choose numbers for the first two squares.
- Fill in the blank squares so that each number is found by adding the two numbers on the left.

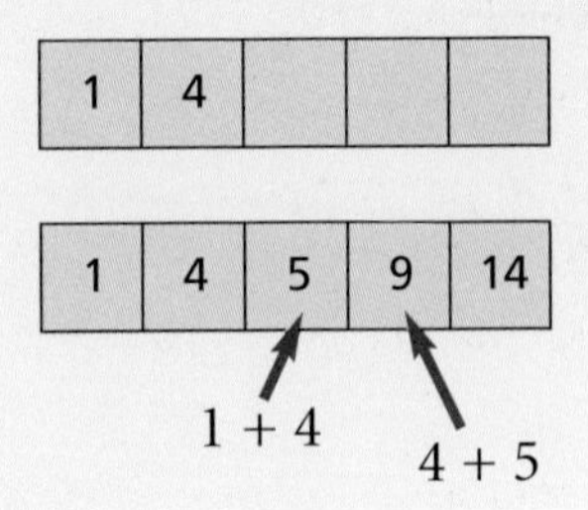

B1 Copy and complete these addition strips.

(a)

⁻1	8			

(b)

5		12	

(c)

9	n	$n+9$		

(d)

⁻3	k		

(e)

	p	4	

(f)

6		m		

B2 (a) Copy and complete this addition strip.

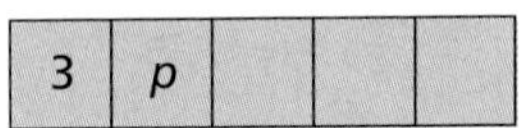

(b) What value of p will give 21 in the last square?

(c) Find the missing numbers in this addition strip.

B3 Find the missing numbers in these addition strips.

(a) 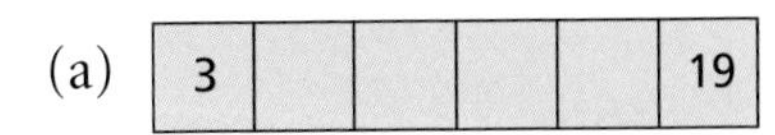

(b) 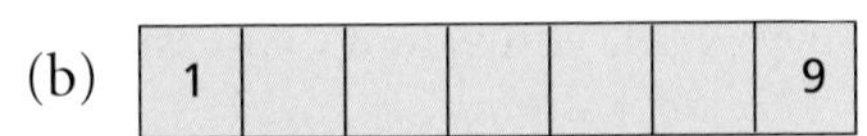

(c) 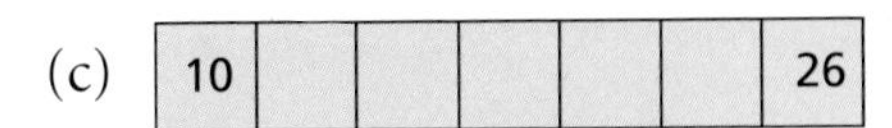

(d) 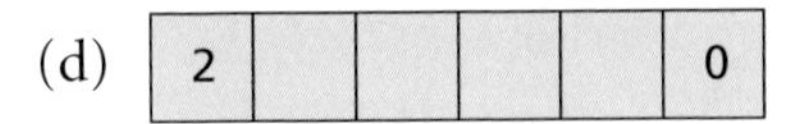

B4 Find the missing numbers in these addition strips.

(a) 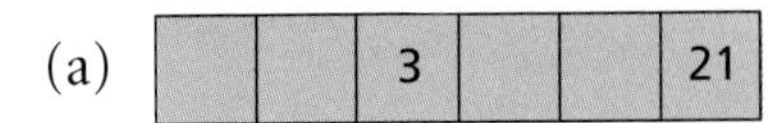

(b) 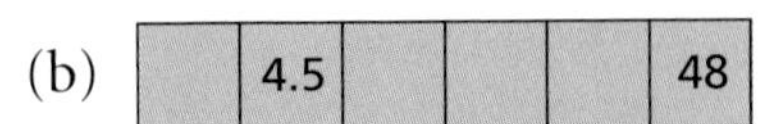

B5 Complete this strip so that the number in the last square is twice the number in the second square.

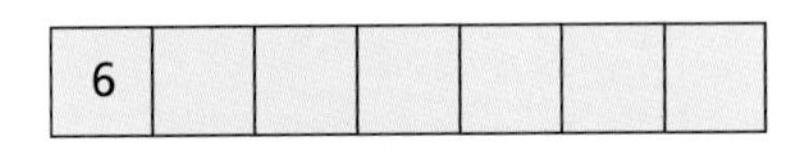

B6 Complete this strip so that the number in the last square is 10 more than the number in the third square.

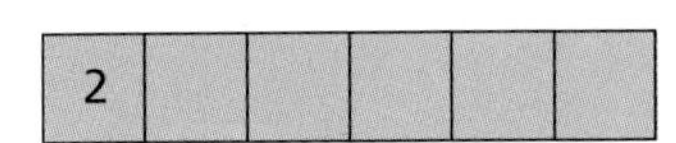

B7 These instructions describe a calculating trick to play on another person.

> **Ask the person to**
> - Make a blank addition strip with six squares.
> - Choose two numbers for the first two squares.
> - Complete the addition strip. (You may need to explain how to do this.)
> - Tell you the number that appears in the 5th square (and nothing else!).

> **You now**
> - Multiply the number by 4 (without showing what you are doing).
> - Ask the other person to add up all the numbers in their strip. Tell them the total will be ... (say the answer to your multiplication).

Try this calculating trick a few times. Use algebra to explain why it works.

C Signs of change?

$10 + (5 + 1)$ $10 + (5 - 1)$ $10 - (5 + 1)$ $10 - (5 - 1)$

- What is the result of each calculation?
- What if there are no brackets?

A
$3q + 11 + (q - 8)$
$= 3q + 11 + q - 8$
$= 4q + 3$

B
$5x - (5 + 3x)$
$= 5x - 5 - 3x$
$= 2x - 5$

C
$7a - (8b - 5a)$
$= 7a - 8b + 5a$
$= 12a - 8b$

C1 Find four pairs of equivalent expressions.

A $2 + (3a - 8)$ **B** $4a - (3a + 6)$ **C** $2a + 1 + (a - 9)$

D $6a - 8 + (2 - 5a)$ **E** $4a - (6 - 3a)$

F $8a - 7 + (1 - a)$ **G** $2a - (8 - a)$ **H** $5a - (6 + 2a)$

C2 Simplify the following expressions.

(a) $6x + (2x - 3)$ (b) $y + 1 + (2y - 5)$ (c) $5 - (z - 3)$

(d) $4x + (6 - 3x)$ (e) $3y - 6 - (y + 3)$ (f) $2z - (z + 7)$

(g) $x + 5 + (3 - 4x)$ (h) $y - (6 - 5y)$ (i) $z + 2 - (5 - z)$

(j) $6x - (2 + 4x)$ (k) $9 - 3y - (7 + 8y)$ (l) $5z - 6 - (3z - 5)$

(m) $10 - 5x - (2 - 3x)$ (n) $12 - (5y + 9)$ (o) $2t - (5t - 3) - 8$

C3 Simplify the following expressions.

(a) $7a + (2b - 3a)$ (b) $4c + d + (3c - 2d)$ (c) $5e - (4f - 3e)$

(d) $8g + (3h - 4g)$ (e) $7j - 5k - (2j + k)$ (f) $6m - (m + 5n)$

(g) $p + 1 + (4q - 8p)$ (h) $r - (3s - 2r)$ (i) $t + 3u - (6u - t)$

(j) $4 + 7v - (w + 2v)$ (k) $10x - y - (6x + 2y)$ (l) $5z - 4y - (2z - 3y)$

D Subtraction strips

- Start with a strip of squares and choose numbers for the first two squares.

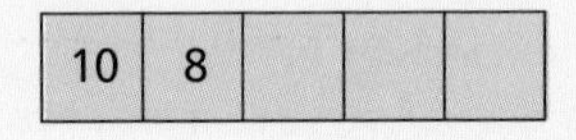

- Fill in the blank squares by subtracting, like this.

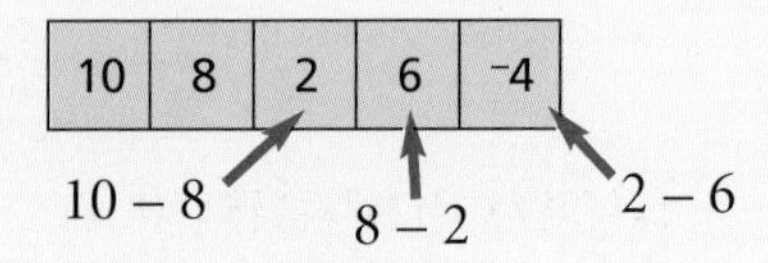

D1 Copy and complete these subtraction strips.

(a)

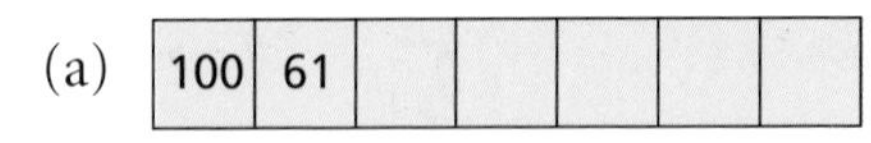

(b)

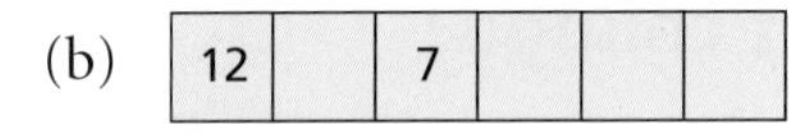

(c)

3	*a*	3–a				

(d)

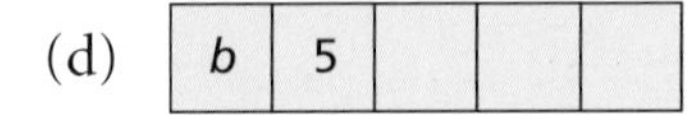

(e)

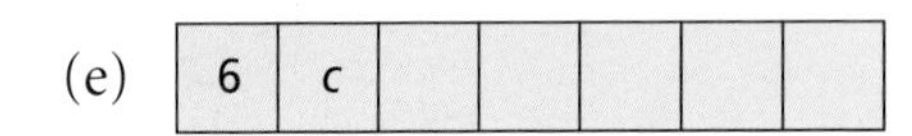

D2 (a) Copy and complete this subtraction strip.

(b) When $n = 15$, what is the number in the last square?

(c) What value of n gives 5 in the last square?

(d) Find n if the number in the last square is 0.

D3 Find the missing numbers in these subtraction strips.

(a)

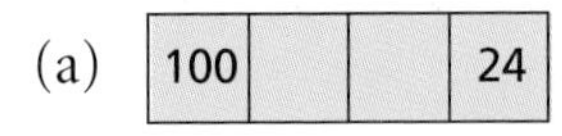

(b)

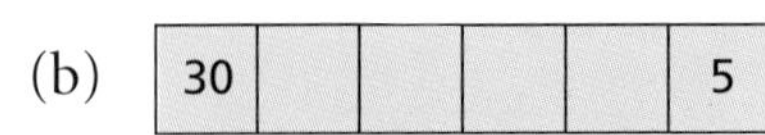

(c)

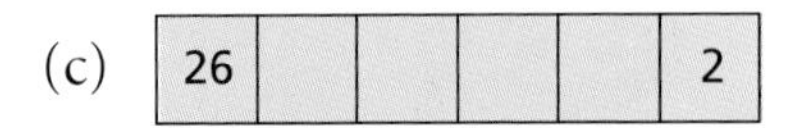

(d)

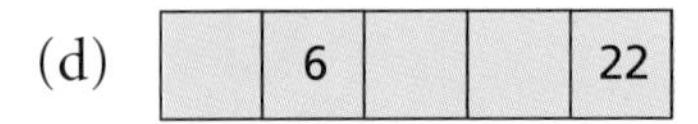

(e)

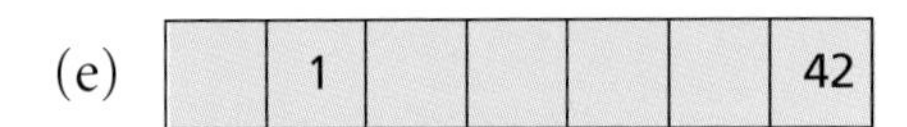

(f)

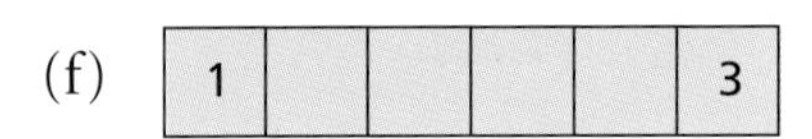

D4 Complete this subtraction strip so that the total of all the numbers in the strip is 60.

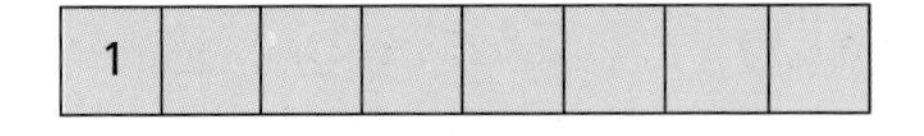

D5

- Ask someone to make a blank subtraction strip with six squares.
  Ask them to choose numbers for the first two squares and complete the strip.
- Then ask them to tell you the number in the second square.
  You multiply this number by 4.
- Ask the other person to add up all the numbers in their strip.
- Tell them that the total will be ... (the number you got by multiplying).

Use algebra to explain why this trick works.

***D6** Find an expression for the length of each coloured piece of this bar. Write your expressions without brackets.

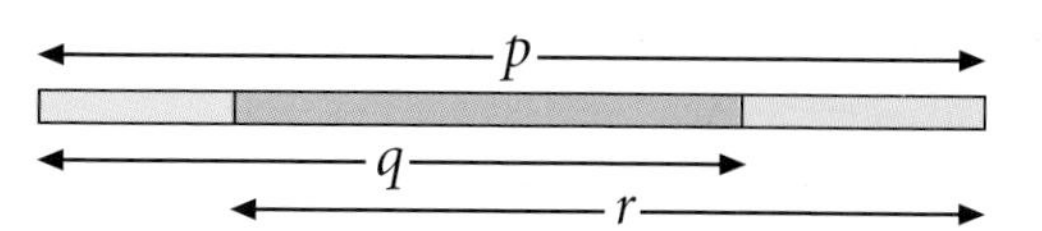

E More simplifying

These show one method of simplifying more complicated expressions.

A
2 + 3p + 3(5 + 2p)
= 2 + 3p + (15 + 6p)
= 2 + 3p + 15 + 6p
= 17 + 9p

B
20y - 4(3y + 5) + 7
= 20y - (12y + 20) + 7
= 20y - 12y - 20 + 7
= 8y - 13

C
7w - 3(w - z)
= 7w - (3w - 3z)
= 7w - 3w + 3z
= 4w + 3z

- How would you describe this method?
- Can you use this method to simplify $6a - 5(a - 1)$?

E1 Find four pairs of equivalent expressions.

A $6p + 2(5 - p)$ **B** $2 + 8p - 4(p - 2)$ **C** $3p - 5(2 - p)$

D $1 + 2p + 3(2p + 3)$ **E** $7p - 3(p + 2)$

F $6(p - 1) + 2p + 16$ **G** $10p - 2(p + 5)$ **H** $p \quad 3(2 - p)$

E2 Simplify the following expressions.

(a) $6x + 5 + 3(x - 2)$ (b) $10y - 3 - 2(y - 5)$ (c) $5z - 10 + 4(6 + z)$
(d) $8w - 5(3 + w)$ (e) $6p - 4(2 - p) + 3$ (f) $13 - 6(q + 1) - q$

E3 Simplify the following expressions.

(a) $5a + 3(2a + 5)$ (b) $8b - 5 - 3(2b - 1)$ (c) $5(c + 1) + 2(3c - 2)$
(d) $2(d + 3) - 4(3d + 1)$ (e) $e - 5(1 - 4e) + 3$ (f) $9f - 2(4f - 3) - f$

E4 Simplify the following expressions.

(a) $8a + 3(2a - b) + 4b$ (b) $2c - 1 - 3(d - c)$ (c) $4(e + 2f) + 3(2e - f)$
(d) $10g + 13h - 5(2g + 2h)$ (e) $7j - 5(2j - 4k) - 1$ (f) $9m - 8(2n + m) + 5n$

What progress have you made?

Statement	**Evidence**

I can simplify expressions such as $2x + 3y - x + y + 4$.

1 Simplify the following expressions.

(a) $4 - q + 1 + 7q$

(b) $6 + 3r - 5r - 1$

(c) $8t - 3s + t + 7s - 10t$

(d) $3w + 5x - w - 7x + 5$

I can simplify expressions such as $4x - (5 - x)$.

2 Simplify the following expressions.

(a) $10 + (5w - 7)$

(b) $7x - (5 + 3x)$

(c) $30 - (11 - 7y)$

(d) $21 + 9z - (13 - 5z)$

(e) $c + 2d - (5d - 3c)$

I can use algebra to solve problems.

3 (a) Copy and complete this addition strip.

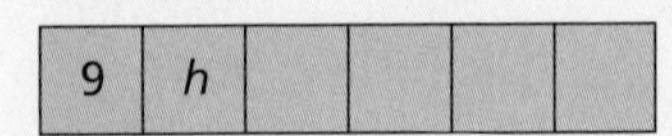

(b) What value of h will give 32 in the last square?

4 Find the missing numbers in these addition strips.

(a)

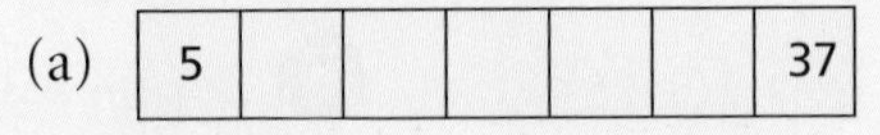

(b)

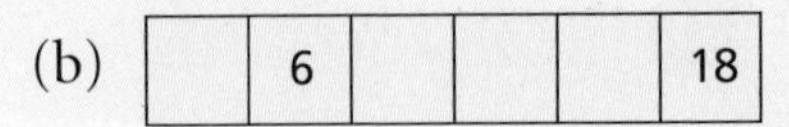

5 Find the missing numbers in these subtraction strips.

(a)

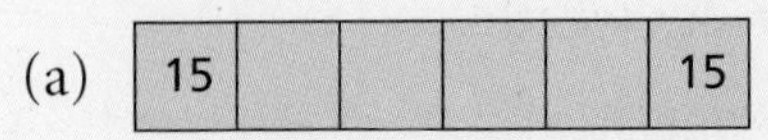

(b)

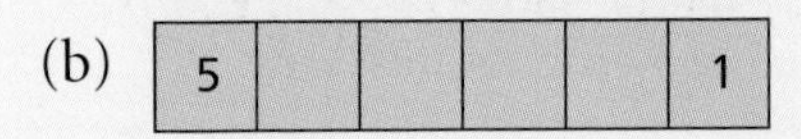

I can simplify expressions such as $16 + 4x - 3(5 - 3x)$.

6 Simplify the following expressions.

(a) $10m - 2(3m - 5)$

(b) $7b - 5c - 2(2b + c)$

20 The right connections

This is about looking for relationships between two sets of data.
The work will help you

- use and draw scatter diagrams and lines of best fit
- understand types of correlation

A 'The Missing Link'

The picture above shows the rock group 'The Missing Link' on the cover of their latest album 'New Connections'.

The **scatter diagram** on the right shows their heights and hair lengths.

- Can you work out which point represents each member of the group?

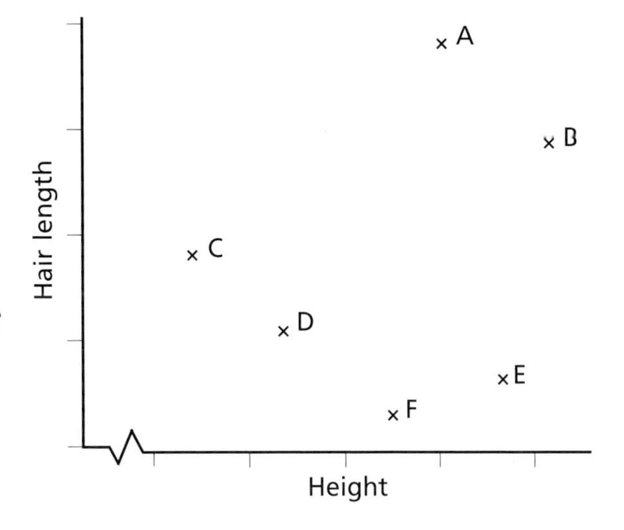

B Scatter diagrams

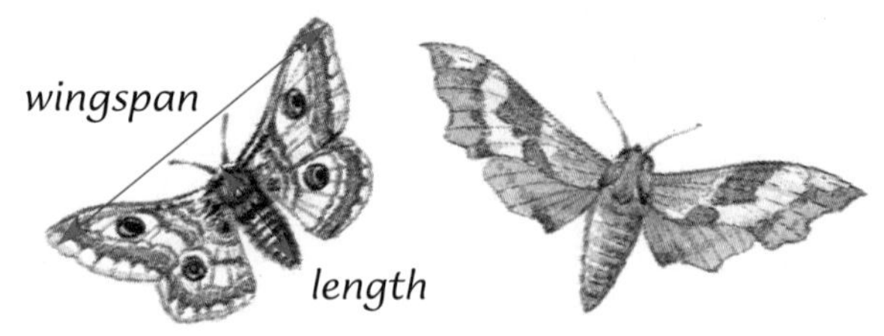

This table shows the body lengths and wingspans of common British moths.

Name	Body length (mm)	Wingspan (mm)
Goat	24	63
Emperor	13	43
Garden Tiger	15	42
Lime Hawk	20	53
Cream Spot Tiger	17	45
Puss	17	50
Peppered	15	46
Wood Tiger	11	29
Yellow Underwing	20	39
Drinker	19	40
Dot	12	29
Scarlet Tiger	10	37
Coxcomb Prominent	13	31
Hook Tip	6	29
Six Spot Burnet	13	26
Ghost	14	40
Hornet Clearwing	19	33
Jersey Tiger	15	48

A scatter diagram of this information has been started below.

The first two moths are already shown.

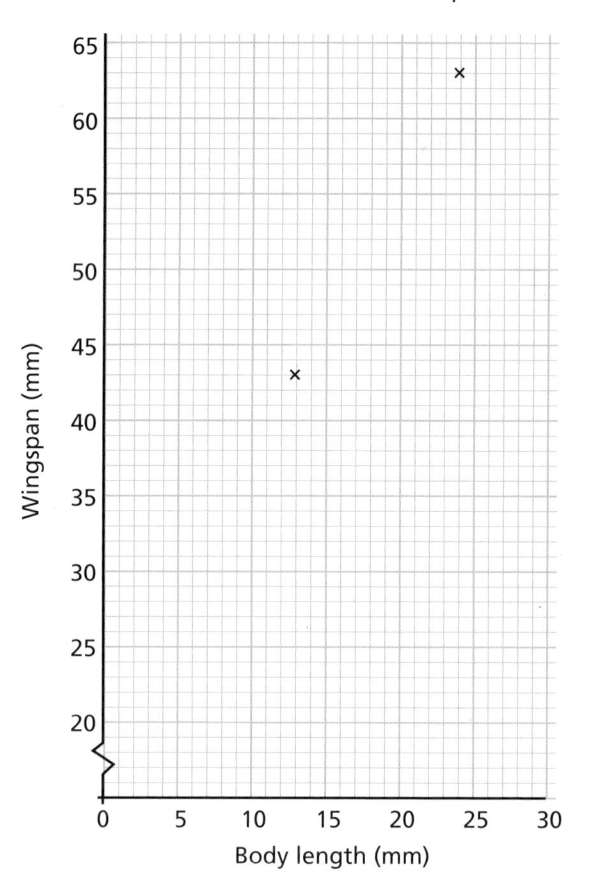

B1 Copy the diagram on squared paper and show all the moths on it.

(a) Does the moth with the longest body have the largest wingspan?

(b) Does the moth with the shortest body have the smallest wingspan?

(c) Do you think it is generally true that moths with longer bodies have larger wingspans?

B2 Some pupils did a 'bleep test' which involves fast running.
Before the test, each pupil's rest pulse rate was measured (in beats per minute).
The results are shown below. (High bleep scores show good performance.)

Pulse at rest (b.p.m.)	60	54	67	69	72	76	59	65	68	90
Bleep score	8.5	11.2	12.0	8.8	9.5	9.3	9.3	10.0	9.4	6.8

Pulse at rest (b.p.m.)	79	88	80	96	82	92	76	83	86	70
Bleep score	5.8	5.3	7.0	3.6	6.5	3.7	6.0	4.3	6.2	7.3

(a) Draw a scatter diagram for the data.
Suitable scales are shown on the right.

(b) What does the diagram suggest about the bleep scores and pulse rates for these pupils?

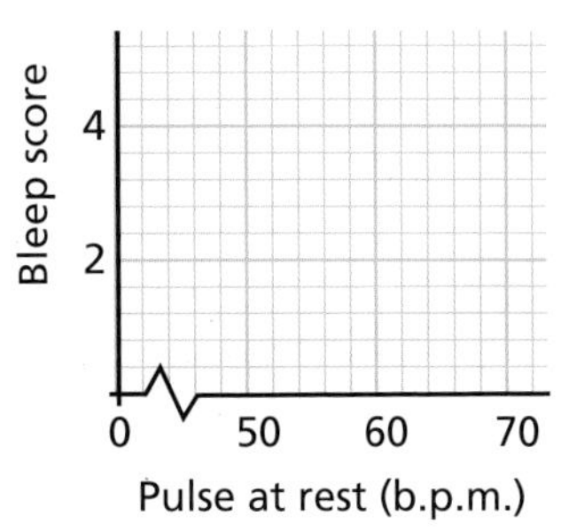

B3 What does each scatter diagram below show?
Write a sentence about each one, for example 'Taller people tend to be heavier'.

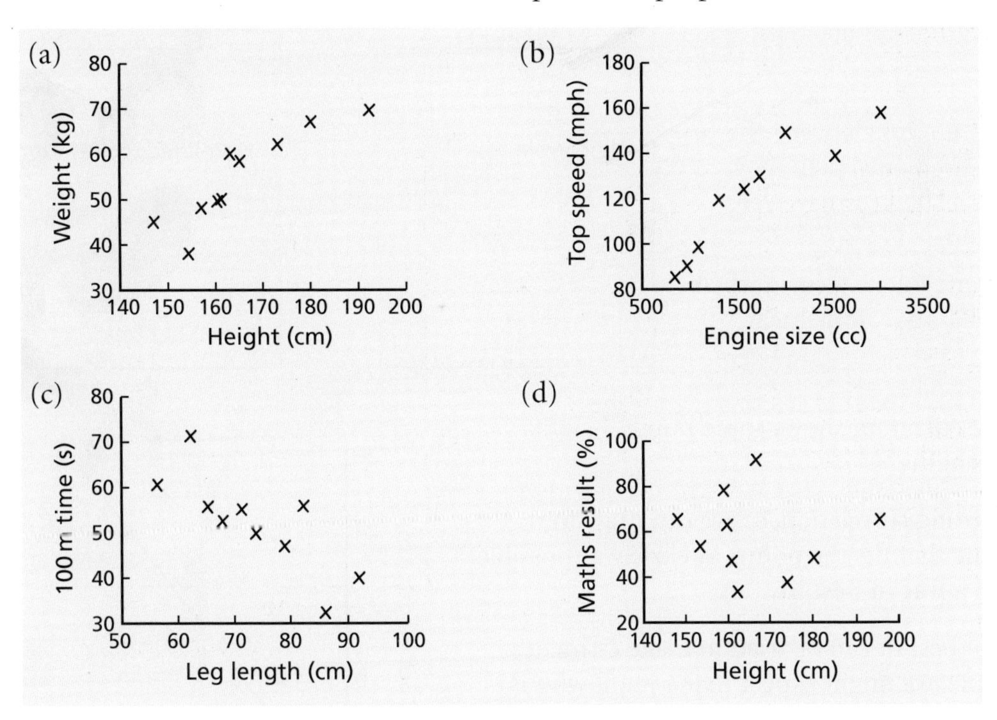

B4 Sketch the scatter diagrams which you think these sets of data would give you.

(a) The time spent by pupils watching TV and the time spent on homework

(b) Hand span and hair length

C Correlation and lines of best fit

If the points of a scatter diagram appear roughly to follow a straight line, then we say that the two measurements are **linearly correlated.**

If the points follow an upward sloping straight line, the correlation is **positive**.

Both measurements increase together.

Strong positive correlation

Weak positive correlation

If they follow a downward sloping straight line, the correlation is **negative**.

One measurement goes down as the other goes up.

Strong negative correlation

Weak negative correlation

If there is no link between the two measurements, then they are uncorrelated (or have **zero correlation**).

If the points follow a curve, then the correlation is **non-linear**.

No correlation

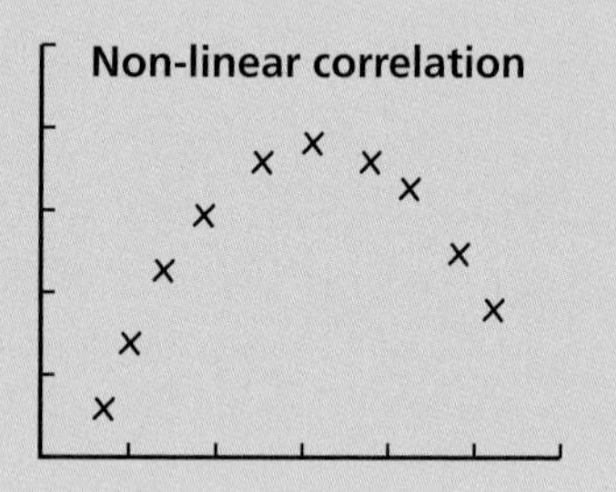

This scatter diagram shows some pupils' heights and foot lengths.

There is strong correlation, so we can draw the straight line that fits the points as closely as possible. It is called a **line of best fit**.

The line of best fit can be used to make estimates. For example, we might expect that a pupil who is 160 cm tall might have a foot length of about 21.5 cm.

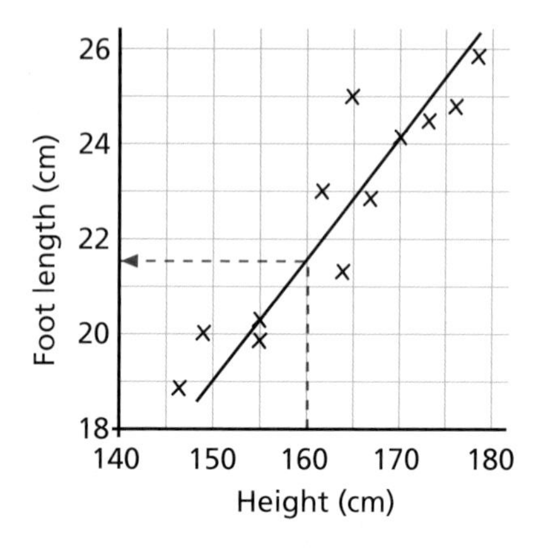

C1 Use the line of best fit to estimate the foot length of a pupil of height 175 cm.

C2 How would you describe the type of correlation shown by each diagram in question B3?

C3 This table gives information about the engine sizes and horsepowers of some cars.

Engine size (litres)	1.8	2.8	3.5	2.9	1.6	1.4	1.3
Horsepower	59	193	202	191	100	75	62

Engine size (litres)	1.3	3.2	1.7	1.5	1.9	2.0	1.2
Horsepower	74	216	130	75	140	115	44

The smallest engine size is 1.2 and the largest 3.5.
The lowest horsepower is 44 and the highest 216.

On graph paper, choose scales for a scatter diagram to make the diagram a reasonable size and roughly square.

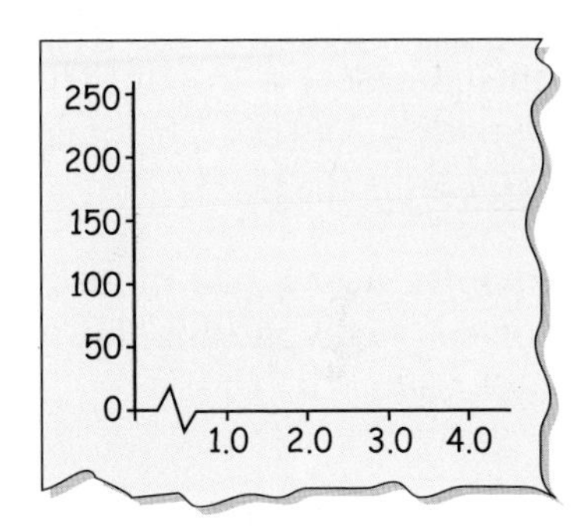

(a) Draw the scatter diagram.

(b) Draw by eye a line of best fit.

(c) Use the line to estimate the horsepower of a car whose engine size is 2.5 litres.

C4 A fish farmer collected this information about salmon.

Weight (kg)	5.0	3.3	4.3	5.1	3.4	5.1	4.7	4.2	3.1	2.9	4.5	4.9	3.4	5.0	5.0
Length (cm)	74	49	62	67	54	74	71	57	54	48	69	69	58	79	71

(a) Plot this information on a scatter diagram and draw a line of best fit.

(b) Measuring the length of a live salmon is difficult.
Weighing is easier, using a net with scales attached.
Use your line of best fit to estimate the length of a salmon weighing

(i) 4 kg (ii) 3 kg (iii) 3.7 kg (iv) 5.5 kg

(c) Estimate the weight of a salmon of length 75 cm.

(d) The largest salmon ever caught was about 1 m long.
Roughly how much would this salmon have weighed?
How reliable do you think your estimate is?

C5 Alex is investigating the price of used cars.
She finds some adverts for the same make and model of car but of different ages in her local paper.

Age (years)	0	2	2	3	4	4	5	5	8	9	10	12
Price (£)	8300	7300	7250	7000	4700	5400	3700	3800	2400	2500	1400	1000

(a) Plot this information on a scatter diagram and draw a line of best fit.

(b) Use your graph to estimate the price of a 7-year-old car.

(c) After how many years roughly should the car have zero value?

D Quarters

Sometimes it is hard to decide whether there is any correlation between two sets of measurements.

This scatter diagram shows some people's heights and footlengths.

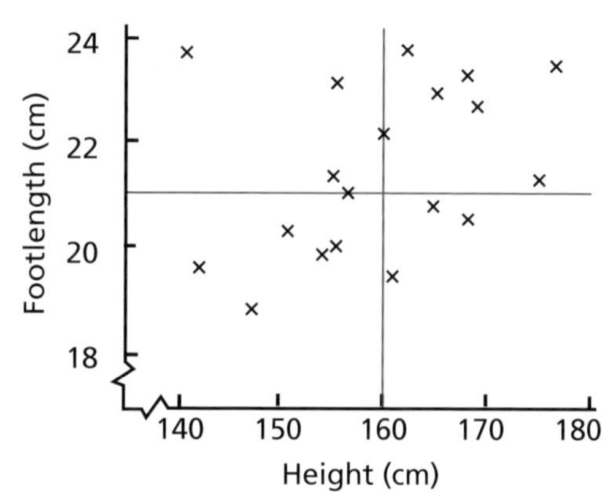

The median height (found by putting the heights in order) is 160 cm. The median footlength is 20.9 cm. The medians are shown on the diagram.

The points in each pair of diagonally opposite 'quarters' can be counted, ignoring those on a line.

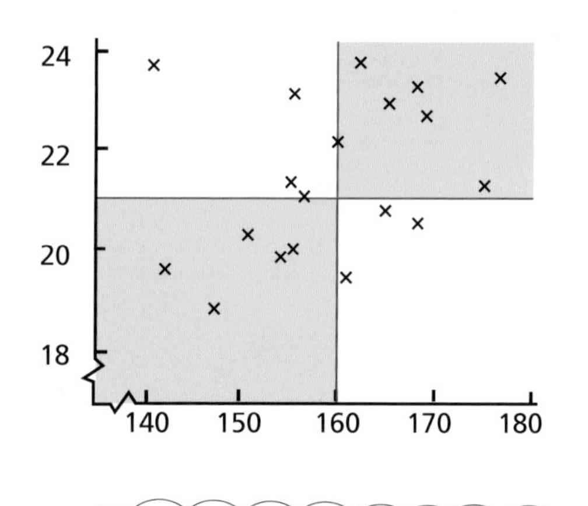

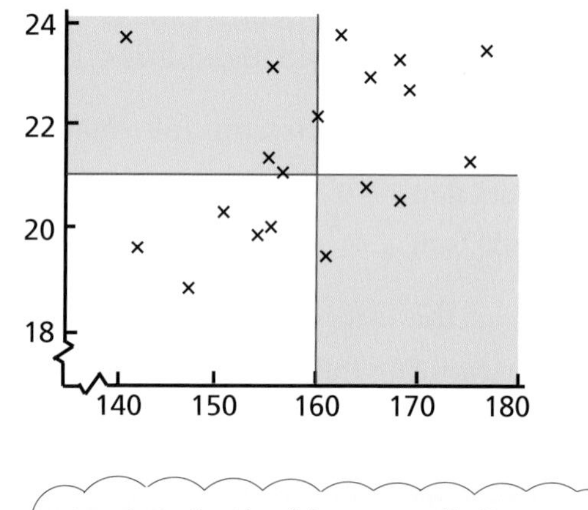

11 points in the 'upward' diagonal

6 points in the 'downward' diagonal

There are many more points in the 'upward' (↗) diagonal than in the 'downward' (↘). This indicates a positive correlation.

D1 What would happen if there were

(a) negative correlation (b) zero correlation

D2 This table shows the mass of carbon monoxide and of nitrogen dioxide given off by 25 car engines for each kilometre driven.
Use the quartering method to find out if there is any correlation between carbon monoxide and nitrogen dioxide emissions.

Carbon monoxide (g/km)	3.1	9.1	5.3	2.7	3.1	4.5	4.7	7.6	9.1	4.8	7.2	2.5	4.6
Nitrogen dioxide (g/km)	0.8	0.4	0.7	0.8	0.7	0.9	0.7	0.8	0.4	0.8	0.8	0.9	0.9

Carbon monoxide (g/km)	3.2	2.5	3.7	1.3	3.9	9.2	2.9	6.0	9.3	11.8	2.6	1.4
Nitrogen dioxide (g/km)	0.7	1.2	0.6	1.1	0.9	0.7	1.1	0.4	0.3	0.5	0.7	1.2

E Are you fit then?

The tests on this page are designed to give you an idea of your fitness.

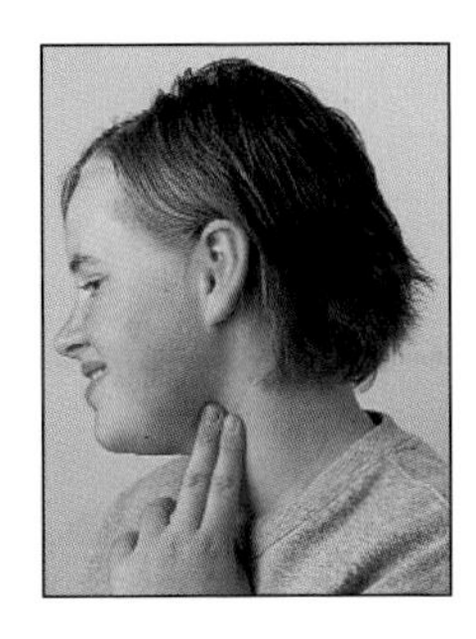

Resting pulse

Before you start any of these tests, sit down for at least two minutes and then take your pulse. The easiest place to take this is at the side of your throat.

Shuttle runs

These are a good test of speed and stamina. The run is made with cones placed 4.5 m apart. Time how long it takes to complete the course.

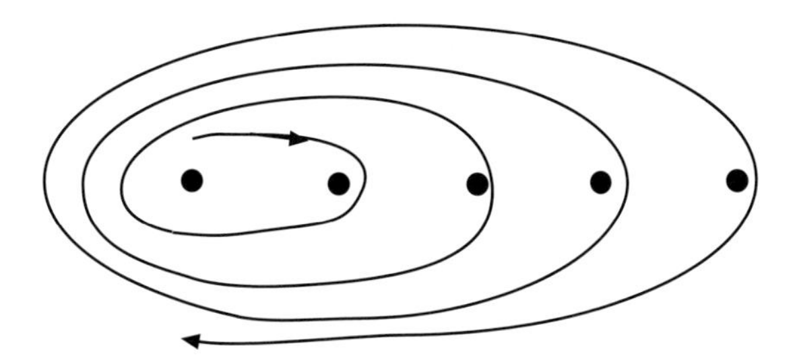

Step-ups

Set up a safe bench about 40 cm high. A 'step-up' consists of stepping up and down again one leg at a time. It helps if another person counts 1-2-3-4 as you do it.

How many can you complete in a minute?

Reactions

Reaction rulers are an easy way to measure reaction times. A partner releases this and you try to catch it.

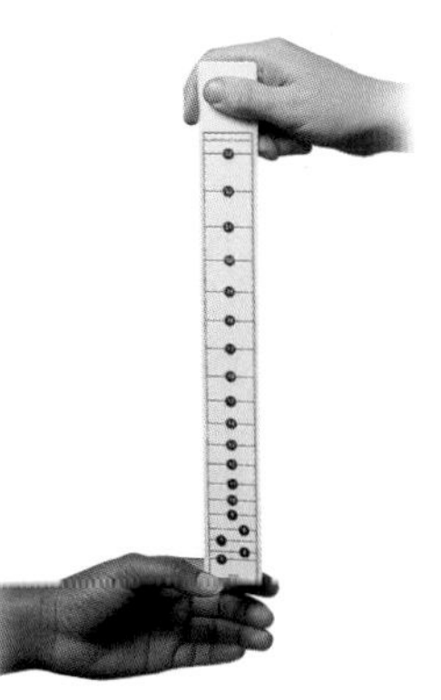

Holding power

Hold a bag at arm's length.
How long can you hold it for?

Record people's height, leg length or anything else which you think might affect their performance in the tests.

Draw scatter diagrams to investigate some hypotheses, such as

'People with longer legs do better at the shuttle run.'

What progress have you made?

Statement	Evidence
I can draw scatter diagrams from data.	**1** Draw a scatter diagram for this data.

Height (cm)	Jump (m)
163	1.3
168	1.4
147	1.1
140	1.1
192	1.9
156	1.2
180	1.7
177	1.5
158	1.4
160	1.2
174	1.6
172	1.4

Statement	Evidence
I can describe connections between measurements.	**2** Describe the connection between height and distance jumped.
I can tell what type of correlation the data shows.	**3** (a) What type of correlation does this data show? (b) What other types of correlation are there? Sketch a scatter diagram to illustrate each one.
I can draw a line of best fit and use it to estimate values.	**4** Draw a line of best fit and estimate how far a person 2 m tall can jump.

Triangles and polygons

This work will help you
- construct a triangle, given information about it
- understand and use angle properties of a triangle
- understand and use angle properties of polygons, including regular polygons

A Constructing triangles

How much information do you need?

A triangle has three sides and three angles – six pieces of information.
How much do you need to know in order to 'fix' the size and shape of the triangle?

Suppose you are told …

… the lengths of the three sides

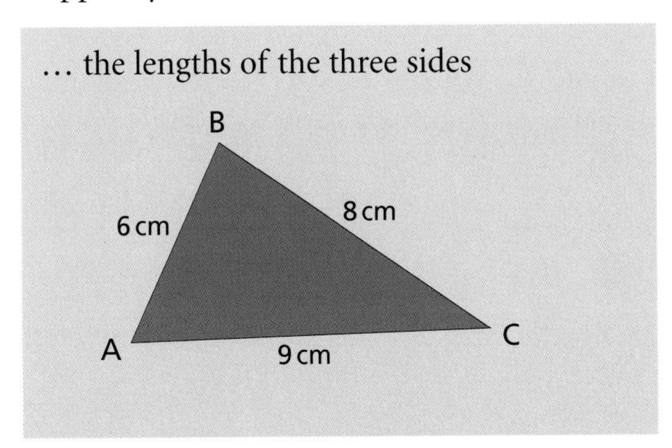

… the sizes of the three angles

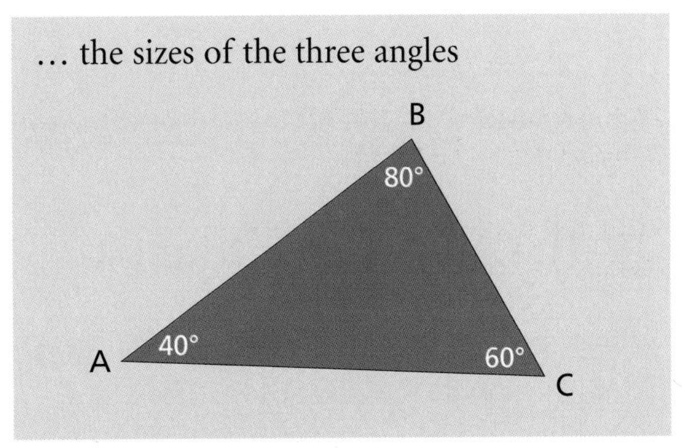

… two sides and the angle between them

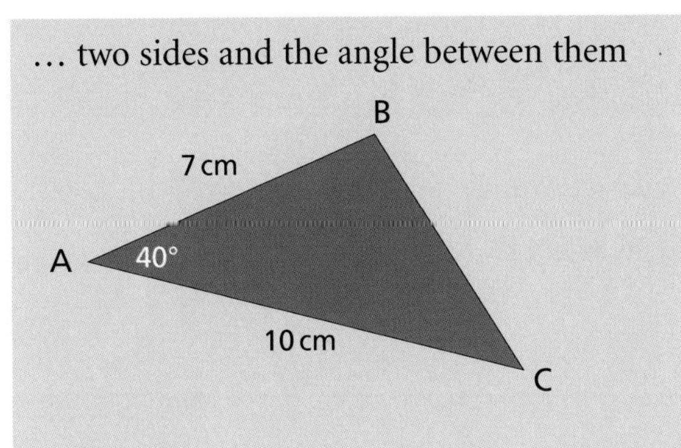

… one side and the angles at each end of it

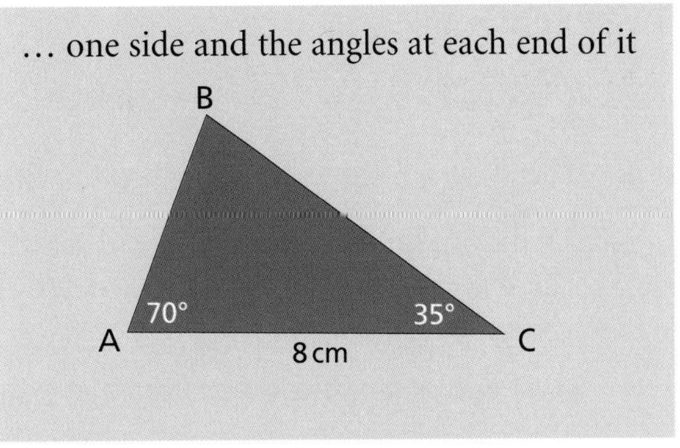

… two sides and an angle other than the one between them

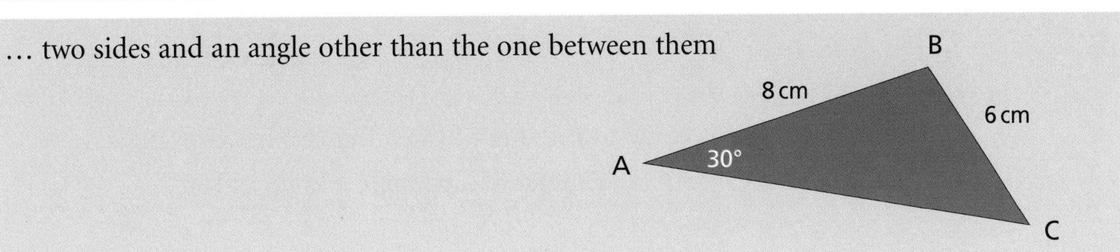

A1 (a) Construct a triangle ABC given that

AB = 7.5 cm   angle A = 40°   angle B = 50°

(b) Measure the lengths AC and BC and the angle C.

A2 (a) Construct a triangle PQR given that

PQ = 5 cm   QR = 8 cm   PR = 9 cm

(b) Measure the angles P, Q and R.

A3 (a) Construct two different triangles XYZ given that

angle X = 20°   XY = 10 cm   YZ = 5 cm

(b) Measure the length of XZ in each triangle.

A4 You have this set of rods (one of each length).

3.5 cm   5.5 cm   7.5 cm   9.5 cm

11.5 cm   13.5 cm

(a) Is it possible to make a triangle with the 3.5 cm, 5.5 cm and 7.5 cm rods as its sides?

(b) Make a list of all the sets of three rods which can be used to make a triangle.

B The angles of a triangle

If one side of a triangle is extended, the angle made is called an **exterior angle** of the triangle.

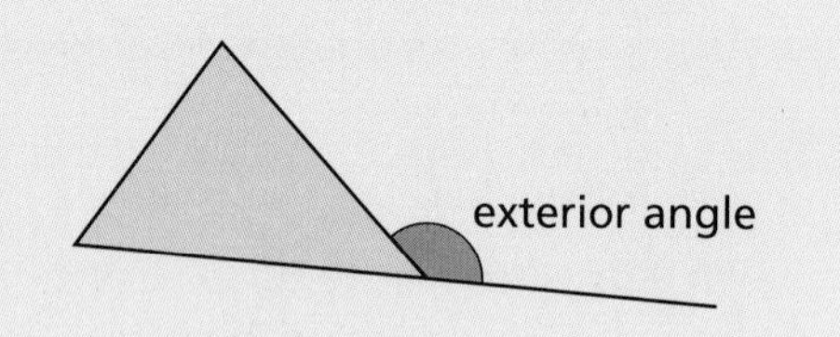

B1 This diagram shows a triangle ABC, with angles a, b and c.
AC has been extended to D, making the exterior angle BCD.
The line CE has been drawn parallel to AB.

(a) Explain why angle ECD is equal to a.

(b) Explain why angle BCE is equal to b.

From (a) and (b) it follows that

the exterior angle is equal to the sum of the other two interior angles.

(c) Explain why $a + b + c = 180°$ (**the angles of a triangle add up to 180°**).

B2 Find the angles marked with small letters.
Explain how you worked out each angle.

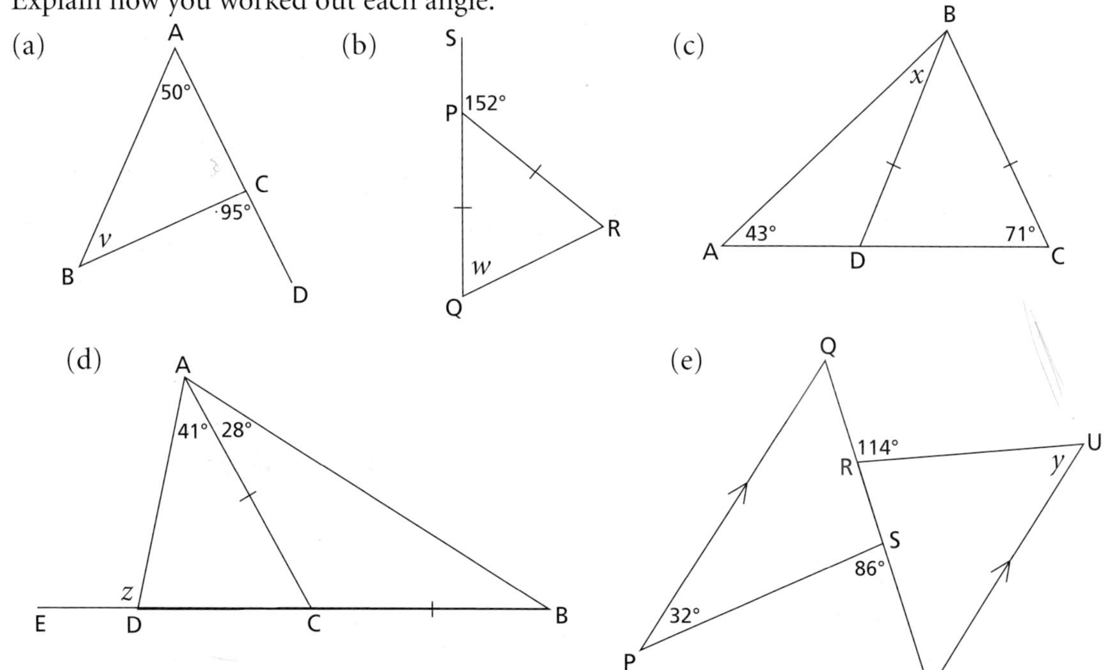

B3 The vertices of this quadrilateral have been joined by straight lines to a point inside the quadrilateral.

(a) Explain why all the lettered angles in the diagram add up to 720°.

(b) What do angles c, f, i and l add up to?

(c) Use parts (a) and (b) to explain why the angles of the quadrilateral add up to 360°.

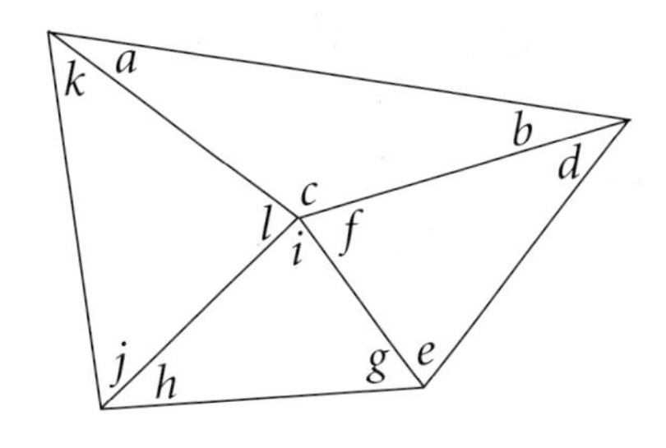

B4 Find the angles marked with letters.
Explain how you worked out each angle.

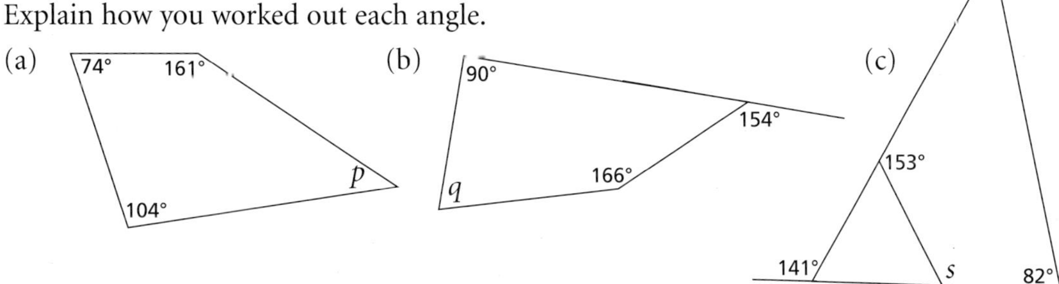

B5 (a) The angles of a triangle are x, $x + 35°$ and $2x + 45°$.
Find the value of x.

(b) The angles of a quadrilateral are y, $y + 40°$, $2y$ and $2y + 50°$.
Find the value of y.

C Interior angles of a polygon

Polygon	Number of sides	Number of triangles	Total of interior angles of polygon
Pentagon	5	3	?

- Find a rule which gives the number of triangles, if you know the number of sides of the polygon.
- How can you work out, in degrees, the total of the interior angles of a polygon?
- What is the total of the interior angles of an n-sided polygon?

C1 What is the total of the interior angles of a polygon with

(a) 12 sides (b) 102 sides

C2 Find the sizes of the angles marked with letters.
Explain your working.

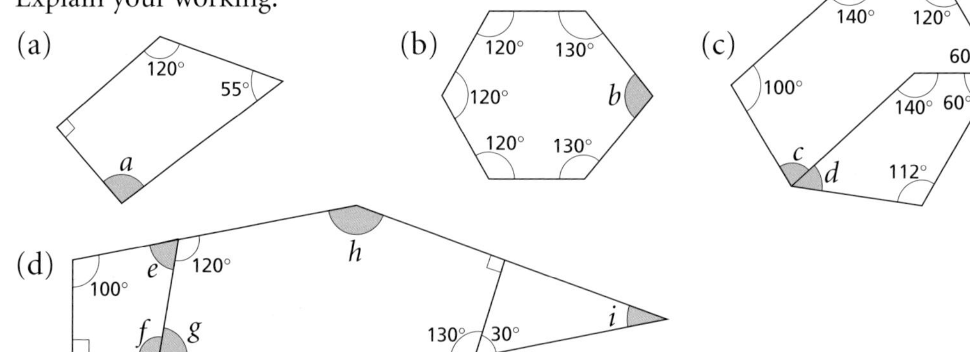

C3 (a) What is the total of the interior angles of a hexagon?

(b) John has drawn this hexagon.
He totals up the interior angles
and says the total is $6 \times 90° = 540°$.
What has he done wrong?

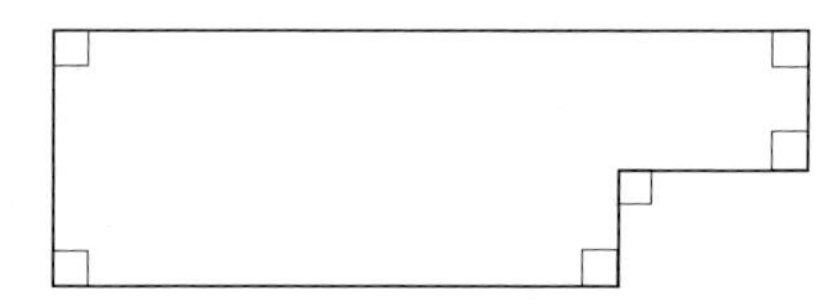

C4 Calculate the size of angle x.

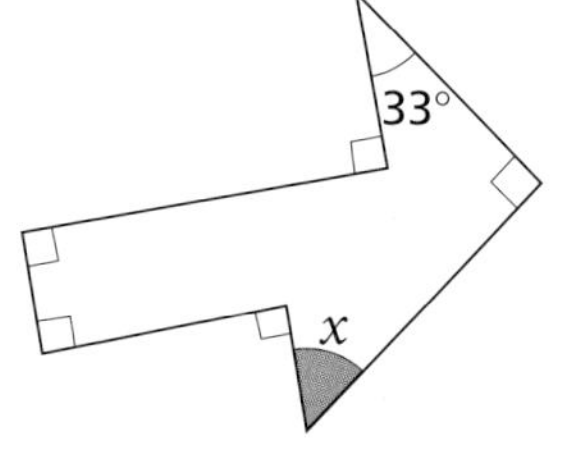

C5 The pentagon has one line of symmetry. Calculate $p + q$.

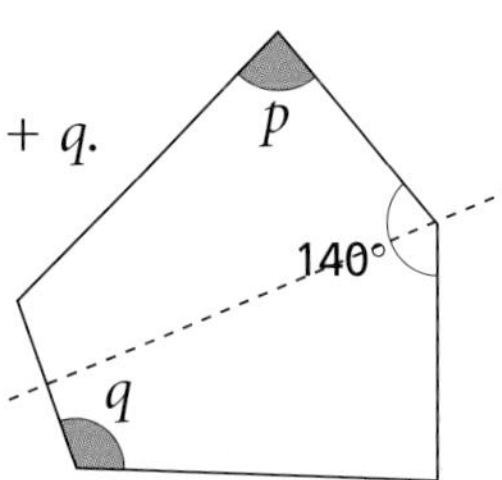

C6 Draw any polygon (with any number of sides).
Mark a point inside the polygon (any point).
Join each vertex of the polygon to the point.

(a) How many triangles have you made?
What is the total of all the angles in all the triangles?

(b) What do the angles at the point add up to?

(c) From your answers to (a) and (b), work out the total of all the interior angles of the polygon.

(d) Do (a) to (c) again, for a polygon with a different number of sides.

(e) Suppose the polygon you start with has n sides.
Go through (a) to (c) again and find a formula for the total of the interior angles.

(f) Use algebra to explain why the formula you have just found says the same as the one found earlier.

C7 The total of the interior angles of a polygon is 2340°.
How many sides does the polygon have?

In a **regular** polygon, all the sides are equal and all the interior angles are equal.

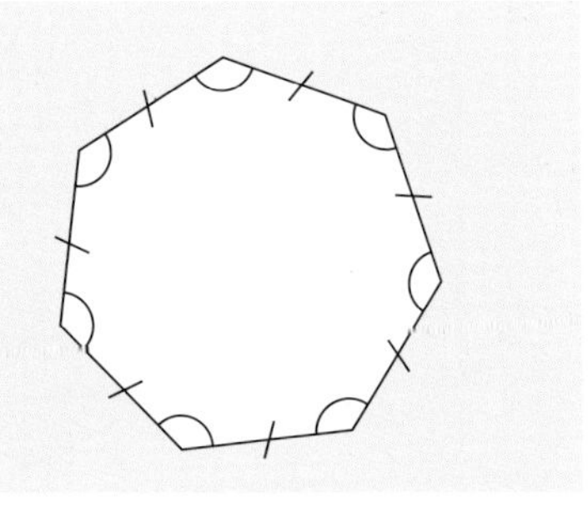

C8 What is the size of each interior angle of

(a) a regular pentagon

(b) a regular octagon

(c) a regular decagon (10 sides)

(d) a regular dodecagon (12 sides)

C9 Calculate the size of each interior angle of a regular heptagon (7 sides).
Give the exact answer, using fractions.

C10 Each interior angle of a regular polygon is 160°.
How many sides does it have?

D Exterior angles

When a side of a polygon is extended, it forms an **exterior angle** of the polygon.

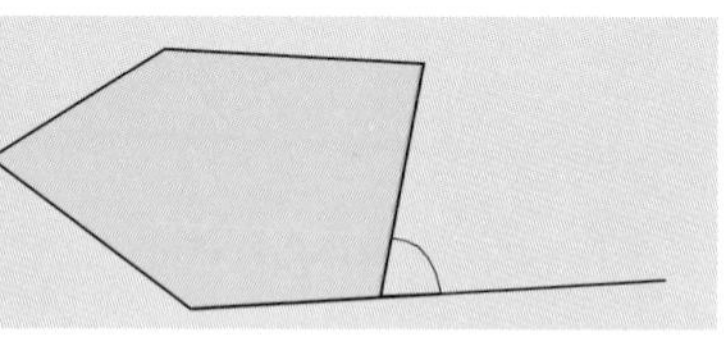

Sliding pieces

- *Draw a polygon (any polygon).*

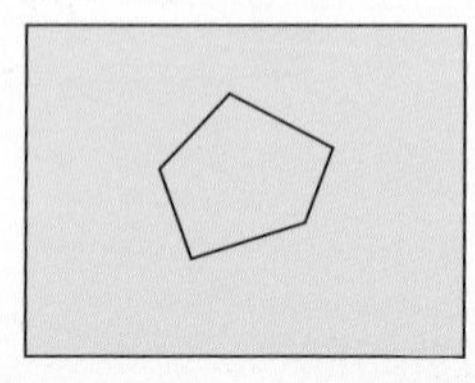

- *Extend the sides of the polygon to the edge of the paper, and mark the exterior angles.*

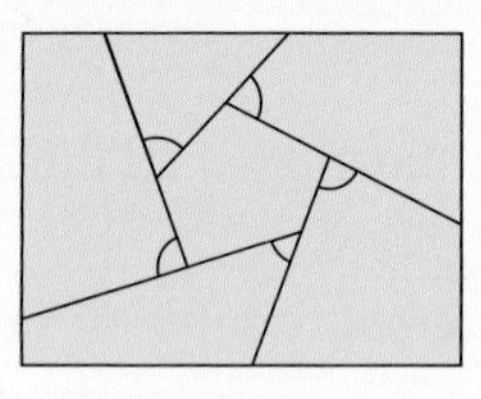

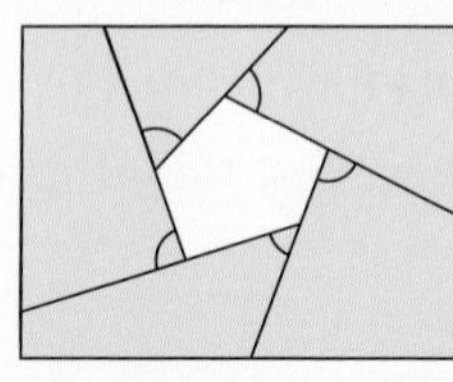

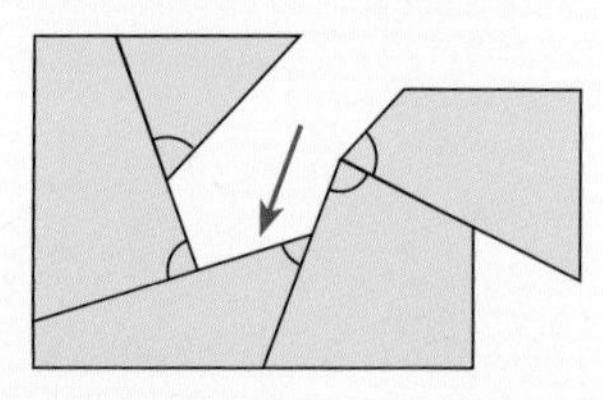

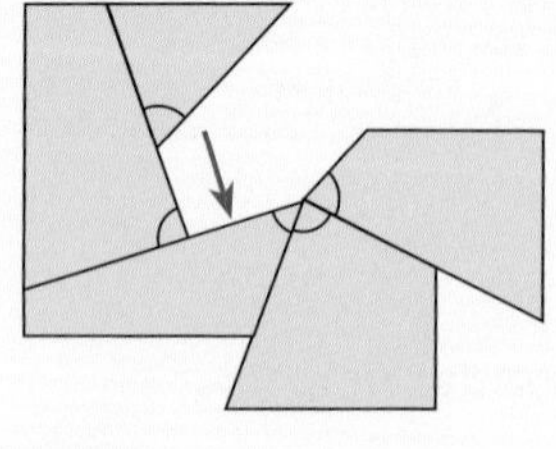

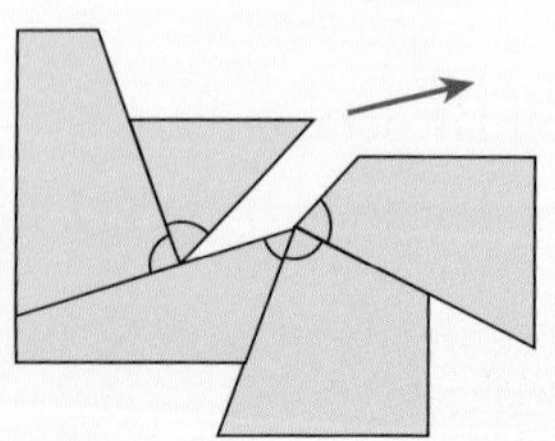

- *What do you discover?*

Pen pushing

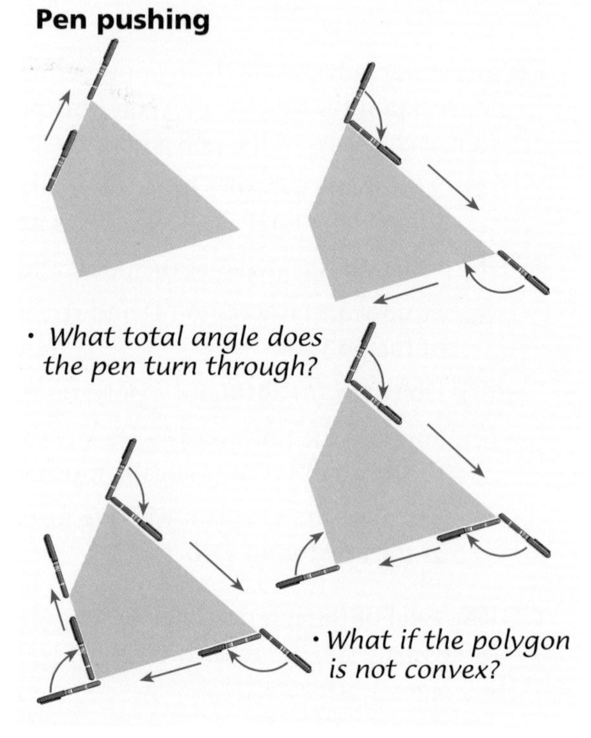

- *What total angle does the pen turn through?*
- *What if the polygon is not convex?*

D1 Write (and simplify) expressions, for an n-sided polygon, for the

(a) total of the interior angles

(b) total of the exterior angles

(c) grand total of the interior and exterior angles

D2 By thinking about the angles at every corner of a polygon,

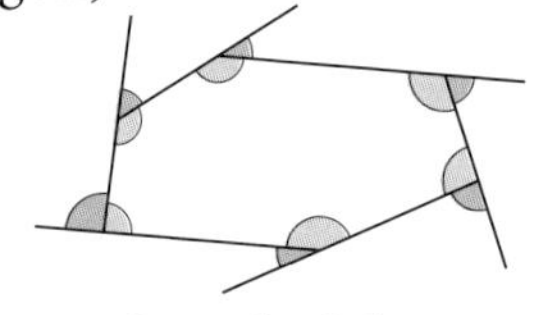

explain why the grand total of the interior and exterior angles of a polygon is what you wrote in D1(c).

D3 Work out the angles marked with letters.

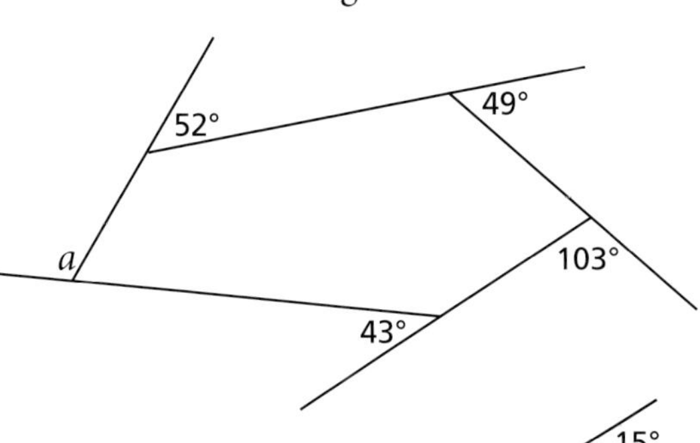

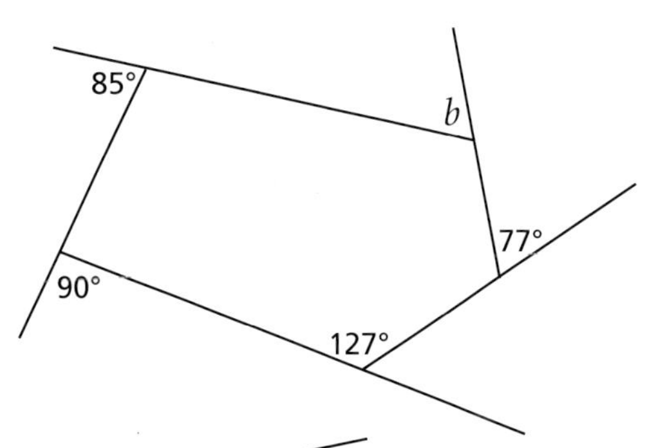

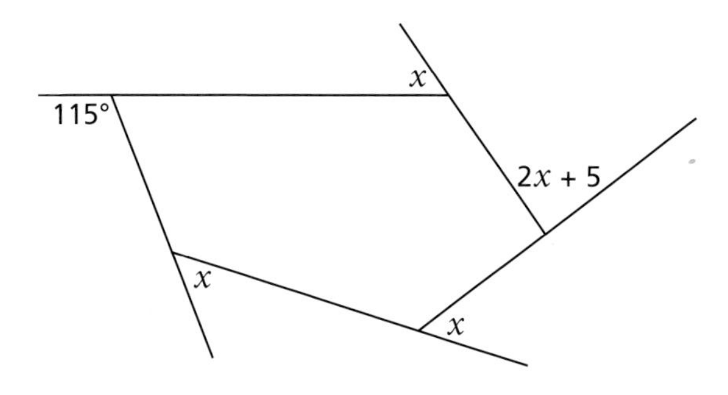

D4 Work out the value of x.

x

115°

$2x + 5$

x

x

D5 In a regular polygon, all the exterior angles are equal.
Calculate the size of each exterior angle of

(a) a regular pentagon
(b) a regular octagon
(c) a regular nonagon (9 sides)
(d) a regular decagon

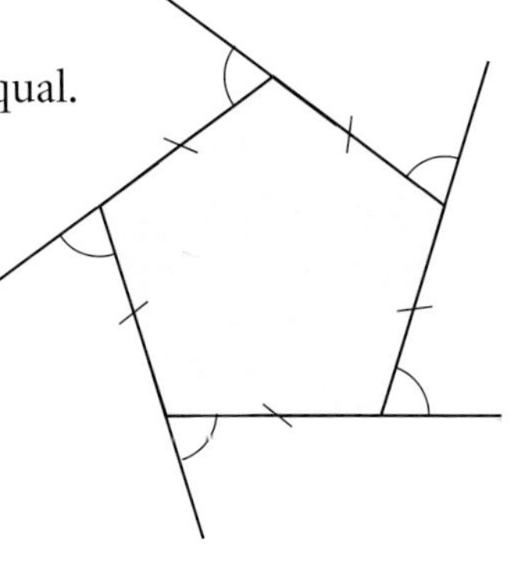

D6 Once you have found the exterior angle of a regular polygon, you can use it to find the interior angle.

Explain how, and use the method to find the interior angle of each regular polygon in D5.

D7 Calculate the size of

(a) each exterior angle of a regular polygon with 20 sides
(b) each interior angle of a regular polygon with 20 sides
(c) each interior angle of a regular polygon with 15 sides

D8 What can you say about the sizes of the exterior angles and the angles at the centre of regular polygons?

Give reasons for your answer.

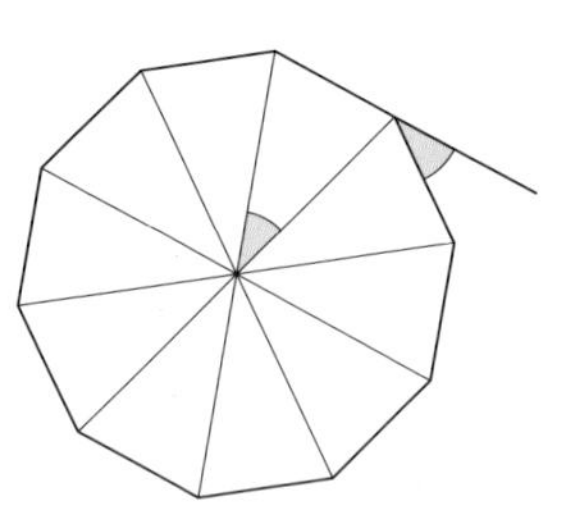

D9 The diagram shows a regular decagon. Calculate the size of angle x.

Explain your method clearly.

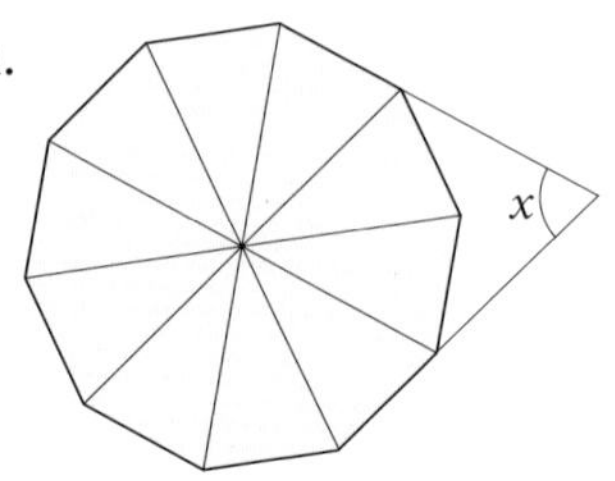

D10 The diagram shows three regular polygons that fit round the point P.
Explain why you can be sure that there is no gap or overlap.

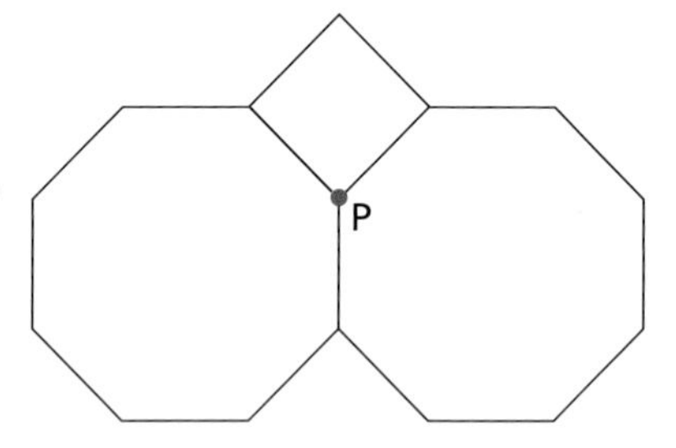

D11 This diagram shows three regular polygons fitting round a point.

Work out the number of sides in polygon A, explaining how you do it.

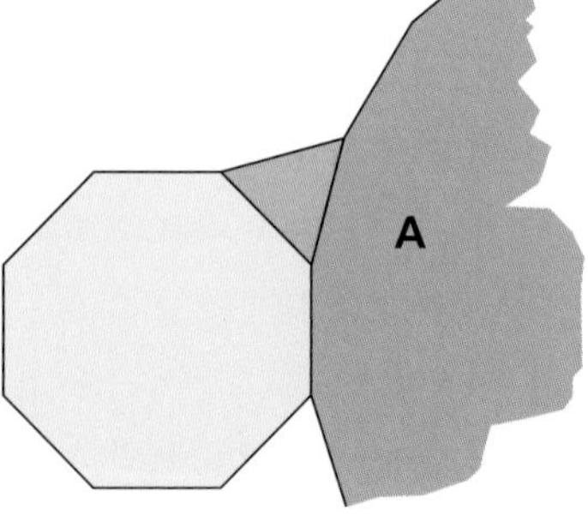

D12 A regular polygon has n sides.

(a) Write down an expression for the size of each exterior angle.

(b) Write down an expression for the size of each interior angle.

D13 An equilateral triangle and two regular polygons fit round a point.

(a) If the two regular polygons are identical, how many sides does each have?

(b) If one of the two regular polygons has 9 sides, how many does the other have?

(c) If one of the two regular polygons has 10 sides, how many does the other have?

(d) If one of the two regular polygons has 7 sides, how many does the other have?

Using LOGO

- What does this LOGO program draw?

```
REPEAT 5(FORWARD 100 LEFT 72)
```

- Modify the progam so that it draws a regular decagon (10 sides).

The angle in the program is the exterior angle of the regular polygon. It is found by dividing 360° by n, where n is the number of sides.

What happens if you choose a value of n which is not a whole number, for example $2\frac{1}{2}$?

- Enter this program and describe what it draws.

```
REPEAT 5(FD 100 LT 360/2.5)
```

- Experiment with different values of n. (You will need to alter the number of 'repeats'.)

What progress have you made?

Statement

Evidence

I can construct a triangle from given information.

1 (a) Construct two different triangles ABC given that

angle A = 25°
AB = 9 cm   BC = 6 cm

(b) Measure the length of AC in each triangle.

I know and can use facts about the exterior and interior angles of a triangle.

2 Calculate the angles marked x and y.

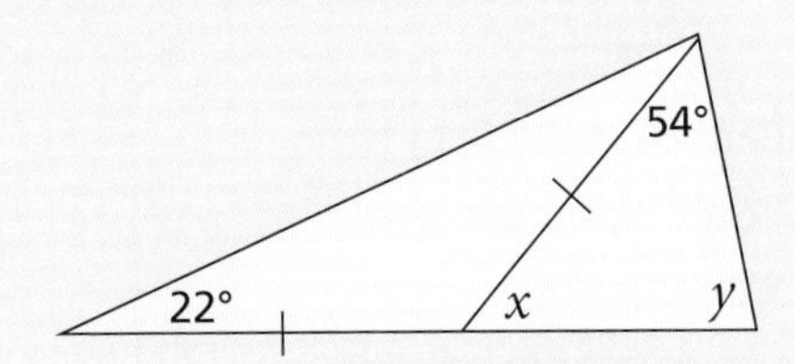

I can solve problems about the interior and exterior angles of polygons.

3 Find the total of all the interior angles of a polygon with 14 sides.

4 A regular polygon has 18 sides. Calculate

(a) the size of each exterior angle

(b) the size of each interior angle

5 Each interior angle of a regular polygon is 168°. Calculate

(a) the exterior angle

(b) the number of sides

22 Moving around

This is about speed.
The work will help you

- understand the connection between distance, time and speed
- calculate and compare speeds
- calculate distances and times

A How fast?

How fast do you walk?

- How could you measure your walking speed?
- If you know your walking speed in metres per minute, you can use this conversion graph to change it to miles per hour (m.p.h.).

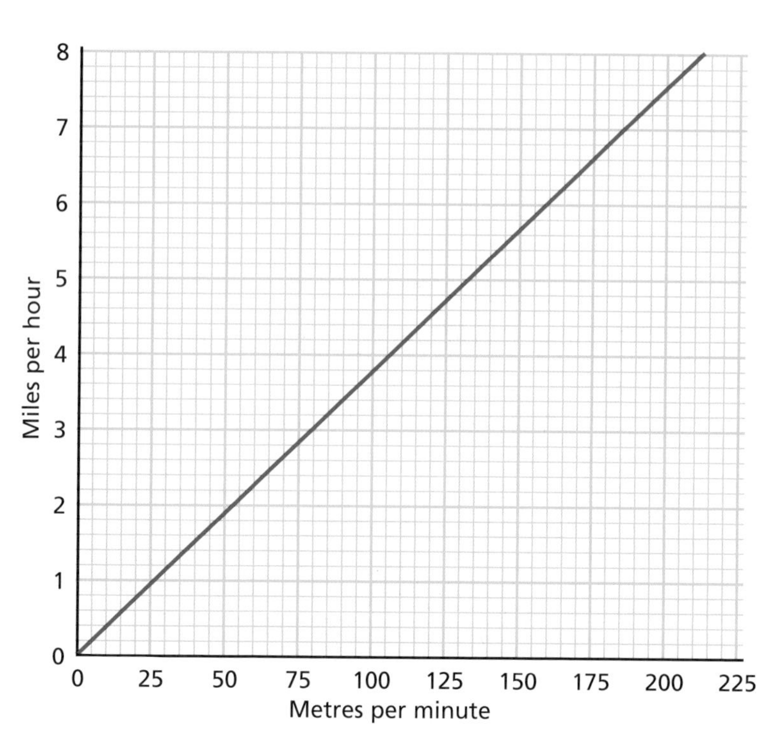

Using a map

You need sheet 207.

The map on sheet 207 shows part of Bournemouth, a seaside town.

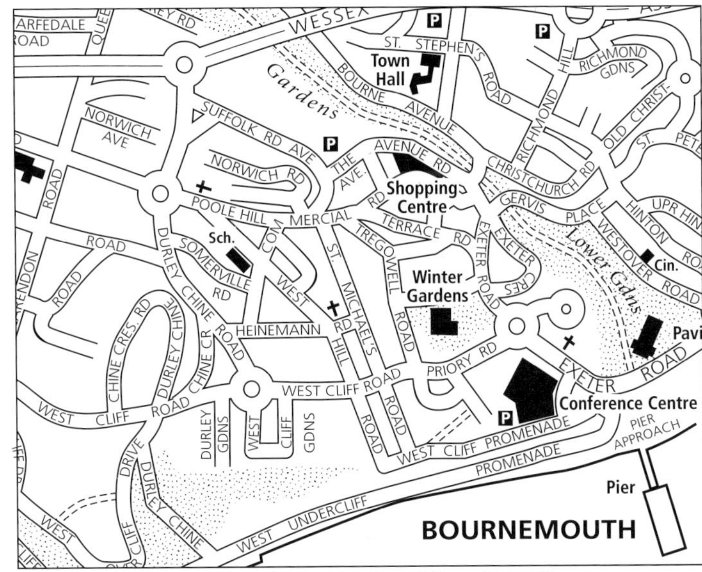

A1 Find the jetty marked in square D5.

(a) How far is it, in metres, along the promenade to the pier in square G4?

(b) How long would it take you to walk this distance?

A2 Find Marlborough Road in squares D3 and E3.
Roughly how long is this road?

A3 Read the instructions below.
Follow the route on the map and say roughly how long it will take you to walk it.

Start at the Pier Approach and walk along the West Cliff Promenade.

Turn right into West Hill Road and then left into West Cliff Road.

Go straight ahead at the roundabout.
Follow West Cliff Road and then turn right into Clarendon Road.

Go along Clarendon Road until you get to the Royal Victoria Hospital.

A4 Plan a circular walk that would take you 1 hour, 2 hours or 3 hours.
Choose a good place to start.
Without drawing on the map, write careful instructions that would enable someone else to do the walk.

Turn left at ...   *Keep going until you see ...*

B Distance, time and speed

Pigeons

Six pigeons were released in Coventry at 12 noon.
They arrived home at the times shown.

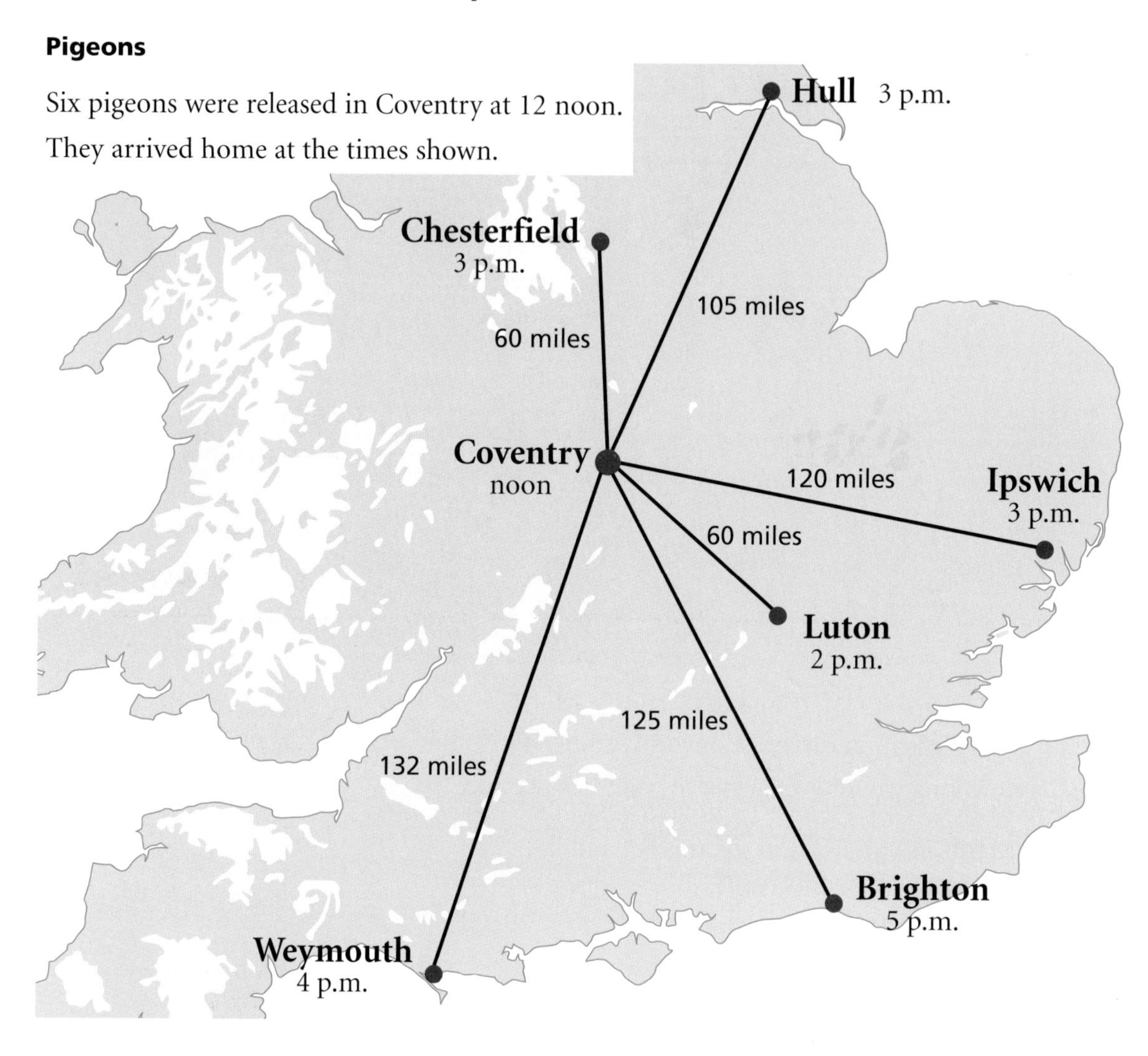

- Which was faster, the Chesterfield pigeon or the Luton pigeon?
- Which was faster, the Hull pigeon or the Ipswich pigeon?
- Put the pigeons in order, from fastest to slowest.

Average speed

The Ipswich pigeon flew 120 miles in 3 hours.
Its **average speed** was 40 miles per hour.

This does not mean that it flew at 40 m.p.h. all the time.
It probably speeded up and slowed down during its flight.

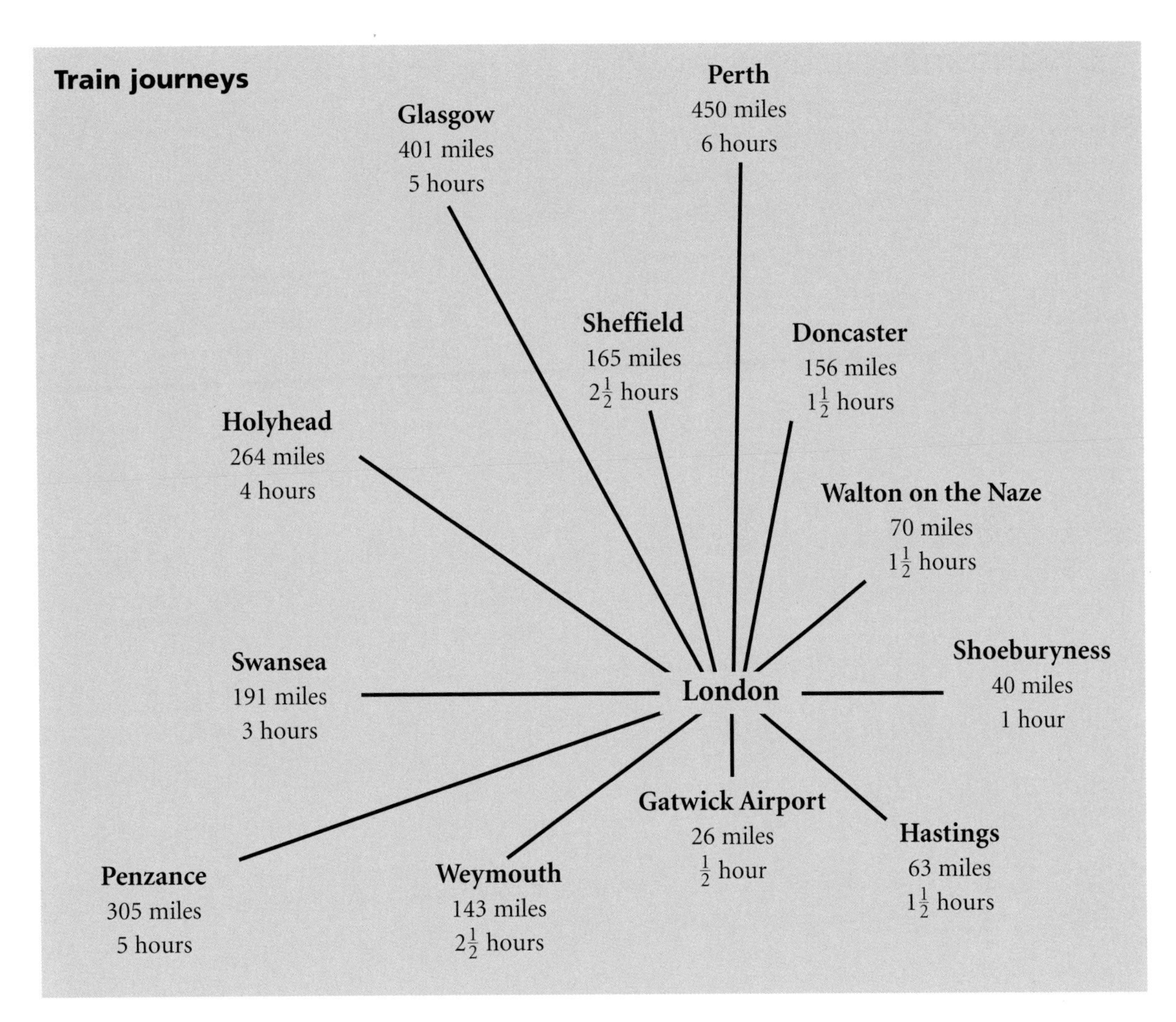
Train journeys
Perth
450 miles
6 hours
Glasgow
401 miles
5 hours
Sheffield
165 miles
2½ hours
Doncaster
156 miles
1½ hours
Holyhead
264 miles
4 hours
Walton on the Naze
70 miles
1½ hours
Swansea
191 miles
3 hours
London
Shoeburyness
40 miles
1 hour
Gatwick Airport
26 miles
½ hour
Hastings
63 miles
1½ hours
Penzance
305 miles
5 hours
Weymouth
143 miles
2½ hours

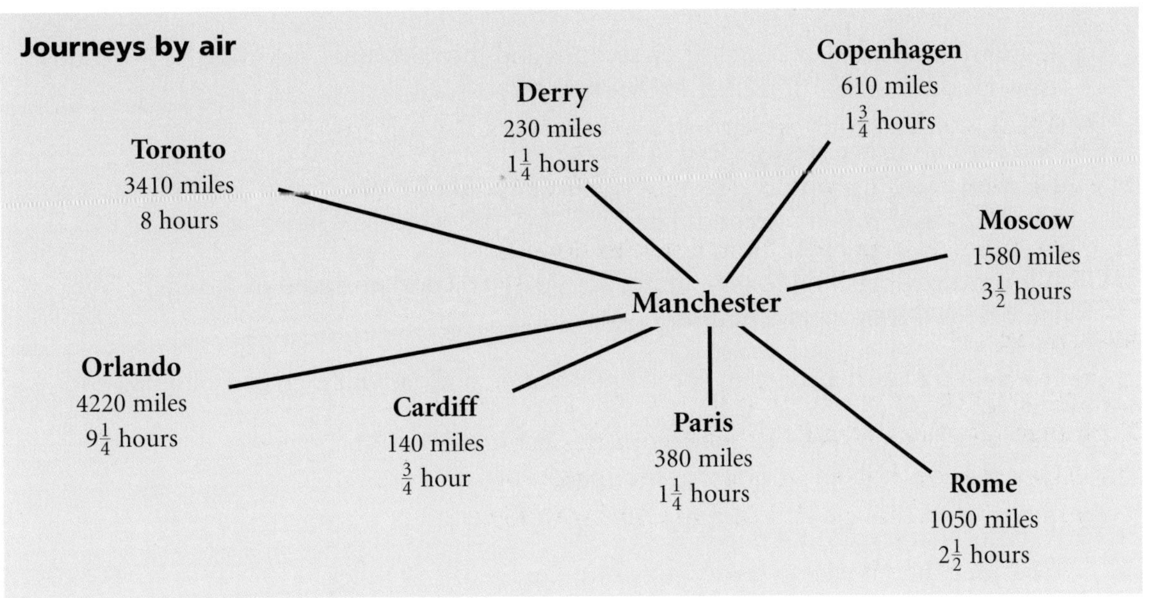
Journeys by air
Copenhagen
610 miles
1¾ hours
Derry
230 miles
1¼ hours
Toronto
3410 miles
8 hours
Moscow
1580 miles
3½ hours
Manchester
Orlando
4220 miles
9¼ hours
Cardiff
140 miles
¾ hour
Paris
380 miles
1¼ hours
Rome
1050 miles
2½ hours

C Constant speeds

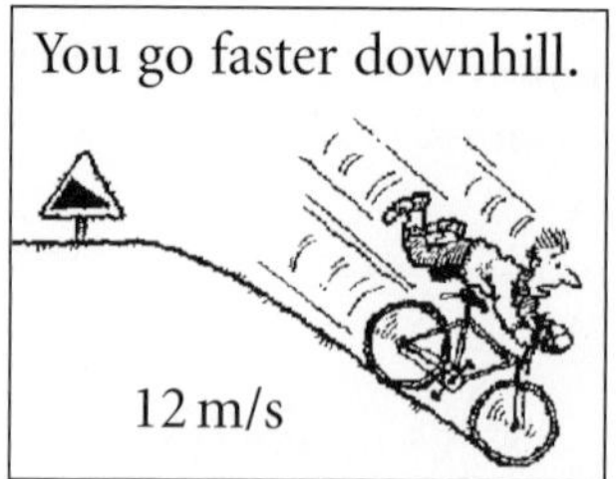

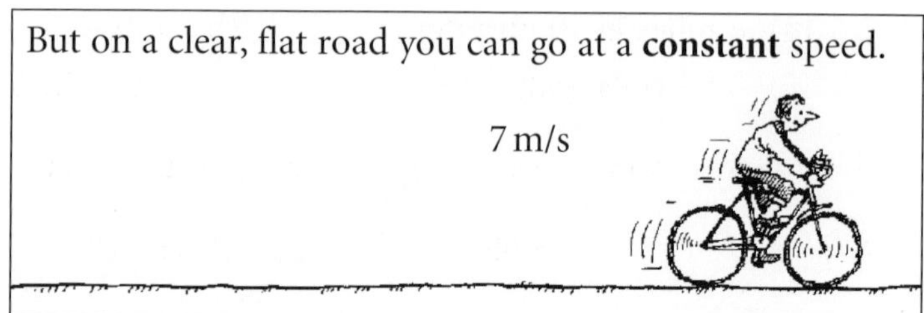

Do not use a calculator for these questions.

C1 How far do you go in 3 seconds at a constant speed of 7 m/s?

C2 If you go at a constant speed of 4 m/s, how far do you go in
(a) 3 seconds (b) 5 seconds (c) 30 seconds (d) 1 minute (e) 5 minutes

C3 An express train is going at a constant speed of 60 m/s.
How far does it go in 8 seconds?

C4 A jet plane is flying at a constant speed of 250 m/s.
How far does it go in (a) 3 seconds (b) 8 seconds

C5 A cyclist goes uphill at 2 m/s for 12 seconds and then downhill at 8 m/s for 15 seconds.
How far does she go altogether in that time?

C6 Bob is running at a steady speed of 4 m/s.
He overtakes Katy, who is walking at a steady speed of 1 m/s.
How far ahead is Bob 5 seconds later?

C7 A lorry travelling at night on a motorway goes at a constant speed of 50 m.p.h.
How far will it travel in 3 hours?

C8 If a train travels at a constant speed of 90 m.p.h., how far will it go in $1\frac{1}{2}$ hours?

C9 A Concorde's top speed is 2300 kilometres per hour (km/h).
How far does it fly in $2\frac{1}{2}$ hours at top speed?

C10 A motorboat travels at 4 m.p.h.
How long does it take to travel (a) 24 miles (b) 60 miles

C11 A coach travels at 50 m.p.h.
How long will it take to travel (a) 200 miles (b) 350 miles (c) 125 miles

C12 A plane flies at 360 m.p.h.
How far does it go in (a) $\frac{1}{2}$ hour (b) $\frac{1}{4}$ hour (c) 1 minute

C13 A train takes 5 minutes to travel 4 miles.
Calculate its speed in miles per hour.

C14 Calculate the speed, in m.p.h., of a train that travels
(a) 6 miles in 10 minutes (b) 4 miles in 6 minutes
(c) 5 miles in 4 minutes (d) 5.5 miles in 5 minutes

C15 This diagram shows the timetable of a train and the distances between stations.

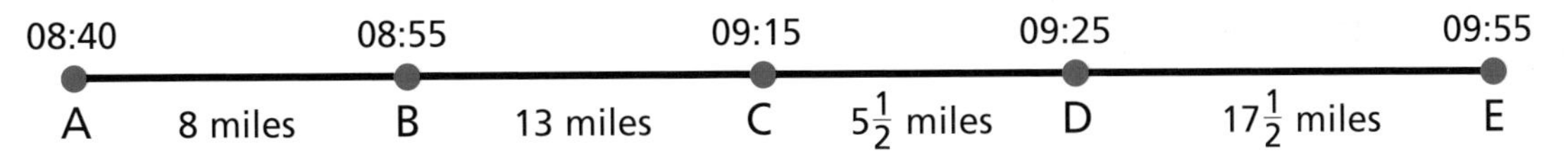

Between which pair of stations is the train (a) fastest (b) slowest
Show all your working.

D Calculating times

Example

A ferry travels at 10 km per hour. How long does it take to travel 35 km?

Imagine the distance the ferry has to travel. 35 km

In each hour it travels 10 km. 1 hour 10 km

In 2 hours it will travel 20 km. 2 hours 20 km

To find the time taken to go 35 km, we need to find how many times 10 goes into 35.

Time taken (hours) = $\frac{\textbf{distance (km)}}{\textbf{speed (km/h)}} = \frac{35}{10} = $ **3.5 hours**

D1 A ship travels at 20 m.p.h.
How long does it take to travel (a) 60 miles (b) 50 miles (c) 10 miles

D2 A train travels at 80 m.p.h.
How long does it take to travel (a) 400 miles (b) 300 miles (c) 20 miles

D3 Use this example to explain how to work out a journey time from the distance and speed.

A plane flies at a speed of 200 m.p.h.
How long will it take to travel 900 miles?

D4 How long will it take a plane flying at 450 m.p.h. to travel 1125 miles?

D5 This diagram shows the route map of an airline. The distances are in kilometres.

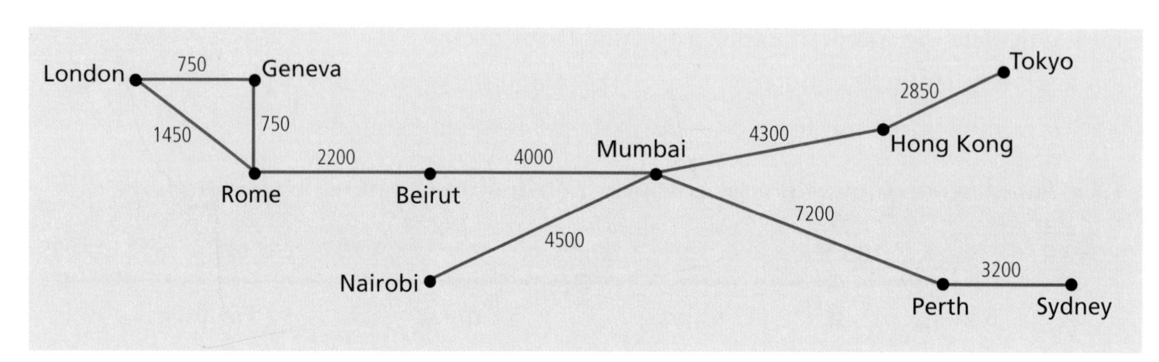

Work out these journey times, to the nearest hour.

(a) Mumbai to Perth at 850 km/h (b) Rome to Beirut at 610 km/h

(c) London to Geneva at 820 km/h (d) Tokyo to Hong Kong at 480 km/h

D6 Chris is working out how long it will take a car to cover 130 miles at a speed of 40 m.p.h.

What is Chris's mistake?

time = distance ÷ speed
= 130 ÷ 40
= 3.25
= 3 hours 25 minutes ✗ *Think again!*

E Time on a calculator

A calculator will show time in decimals of an hour, not hours and minutes.

To change decimals of an hour to minutes, multiply by 60.

Example

0.18 hours = 0.18 × 60 minutes
= 10.8 minutes
= **11 minutes** (to the nearest minute)

To change minutes to decimals of an hour, divide by 60.

Example

37 minutes = $\frac{37}{60}$ hour
= 0.61666… hour
= **0.62 hour** (to 2 d.p.)

E1 Change these to minutes.

(a) 0.25 hour (b) 0.2 hour (c) 0.1 hour (d) 0.45 hour (e) 0.95 hour

E2 Change each of these to a decimal of an hour (to 2 d.p.).

(a) 47 minutes (b) 40 minutes (c) 21 minutes (d) 8 minutes (e) 57 minutes

E3 (a) Write 2.32 hours in hours and minutes (to the nearest minute).

(b) Write 4 hours 17 minutes in hours, to two decimal places.

E4 A boat travels at 14 km/h. How long does it take to travel 32 km?
Give your answer in hours and minutes, to the nearest minute.

E5 A plane leaves Manchester at 08:00 and flies to Aberdeen, a distance of 250 miles.
The plane flies at 210 m.p.h. At what time does it arrive?

E6 Sandra's plane flies at 165 m.p.h.
She is travelling to her home airfield, which is 450 miles away.
The time is now 09:00. She wants to be home by 11:30.

Will she get there in time? If not, how late will she be?

F Mixed questions

F1 This scatter diagram shows how far from school some children live, and their journey times to school.

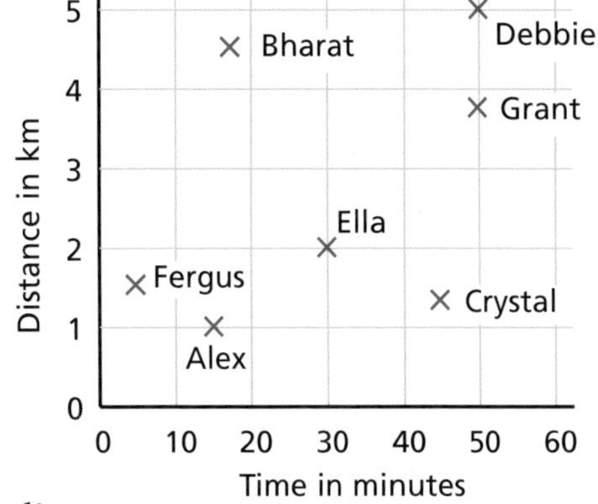

(a) Which of these children

(i) lives furthest from school?

(ii) lives closest?

(iii) takes the shortest time to get to school?

(iv) has the highest average speed?

(v) has the lowest average speed?

(b) Which two children have the same average speed?

F2 Boris ran 800 metres in 1 minute 45 seconds.
Calculate his average speed in metres per second, to one decimal place.

F3 A lifeboat travels at 35 kilometres per hour.
How long does it take to reach a ship in danger 48 kilometres away?
Give your answer in hours and minutes, to the nearest minute.

F4 A plane flies a distance of 254 miles at a speed of 320 m.p.h.
How long, in minutes, does the journey take? Give your answer to the nearest minute.

***F5** Janet drove from London to Brighton (56 miles) and back.
The whole journey, there and back, took 4 hours.
Janet's average speed on the way to Brighton was 42 m.p.h.
What was her average speed on the way back to London?

What progress have you made?

Statement	Evidence
I can calculate an average speed from a distance and a time, and use the correct units for it.	**1** Calculate the average speed of (a) a woman who walks 150 metres in 50 seconds (b) a train which takes 3 hours to travel 111 miles (c) a plane which flies 840 kilometres in 3.5 hours
I can calculate how far something will travel in a given time at a particular speed.	**2** Calculate the distance travelled by (a) a man who jogs for 25 seconds at a speed of 5 metres per second (b) a coach which travels at 45 m.p.h. for 3 hours (c) a plane which flies for 2.5 hours at a speed of 520 km/h
Given a distance and a speed, I can calculate the time taken.	**3** Calculate the time taken by (a) a girl who walks 24 km at a speed of 4 km/h (b) a ship which travels at 14 m.p.h. and covers a distance of 77 miles (c) a plane which flies a distance of 900 miles at a speed of 400 m.p.h.
I can change between minutes and decimals of an hour.	**4** Change 22 minutes to a decimal of an hour, to two decimal places. **5** The distance between two Channel ports is 42 km. How long will the journey take (to the nearest minute) (a) in a ferry travelling at 18 km/h (b) in a hydrofoil travelling at 50 km/h

23 Substitution

This work will help you to substitute values into formulas where several letters occur.

A Letters for numbers

To tell someone how to find the area of a rectangle we could say

Multiply the length by the width.

or we might shorten this to

Area = Length × Width.

Or, if the area is A square units, and the length is L units and the width is W units, we could say

$A = LW$ (where LW means $L \times W$)

This is a **formula** for the area.

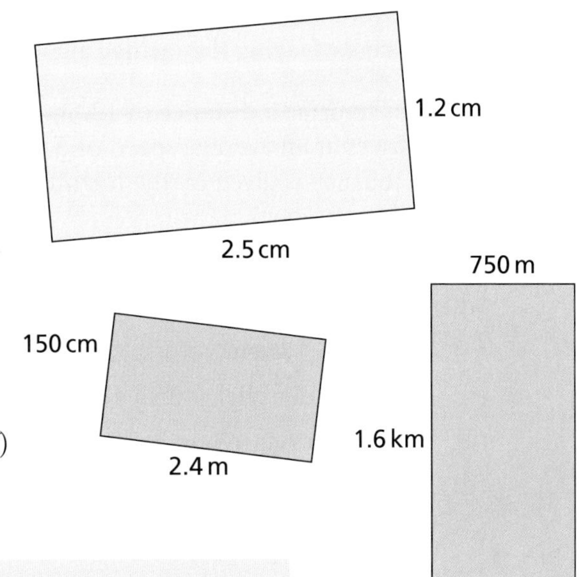

When using formulas

- It must be clear what each letter stands for.
- Think carefully about the units of any quantities.
- Give any answers complete with units.
- Do not give answers more accurate than the original data.

A1 A rectangle has length L and width W. Its perimeter P (the total distance round the edge) is given by the formulas

$P = L + W + L + W$ or $P = 2L + 2W$ or $P = 2(L + W)$

Convince yourself that these are all equivalent formulas.

Use whichever version you find most convenient to find the perimeters of the following rectangles.

(a) A field 120 m long and 200 m across

(b) A poster 40 cm wide and 65 cm high

(c) A postage stamp measuring 18 mm across and 2.5 cm from top to bottom

(d) A doormat measuring 1.2 m by 75 cm

A2 Using the formula $A = LW$, calculate the areas of each of the rectangles in question A1.

A3 A straight tarmac footpath extends from the front door of a house for 7.5 m. It is 80 cm wide and has concrete edging along the sides and each end.

Using the formulas in questions A1 and A2, calculate

(a) the total area of tarmac on the footpath

(b) the total length of concrete edging

(c) the cost of laying the tarmac at £7.25 per square metre

A4 Sarai has to drive a distance of 120 miles.

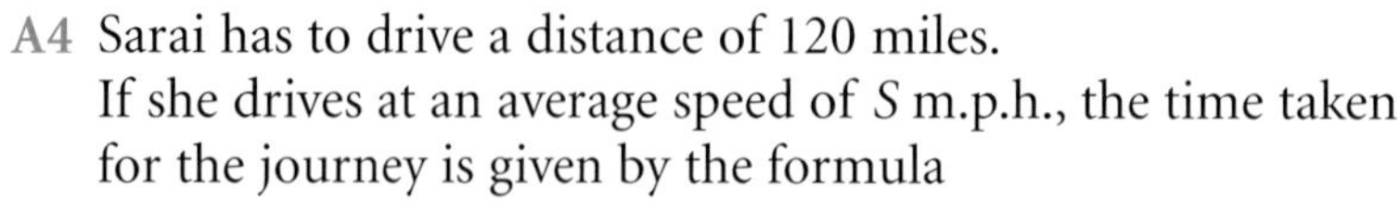

If she drives at an average speed of S m.p.h., the time taken for the journey is given by the formula

$$T = \frac{120}{S}$$

where T is the journey time in hours.

(a) Calculate T when S is 40.

(b) How long does her journey take if she travels at 30 m.p.h.?

(c) What happens to T as S gets larger and larger?

(d) What happens to T as S gets smaller and smaller?

(e) Why is it impossible to calculate T when S is 0?

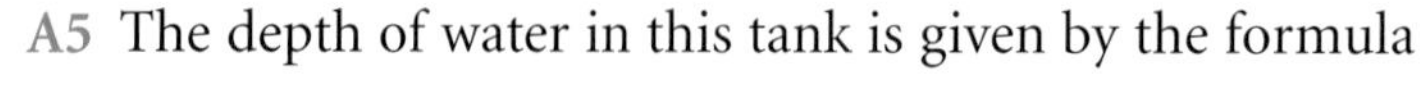

A5 The depth of water in this tank is given by the formula

$$d = 60 - 5t$$

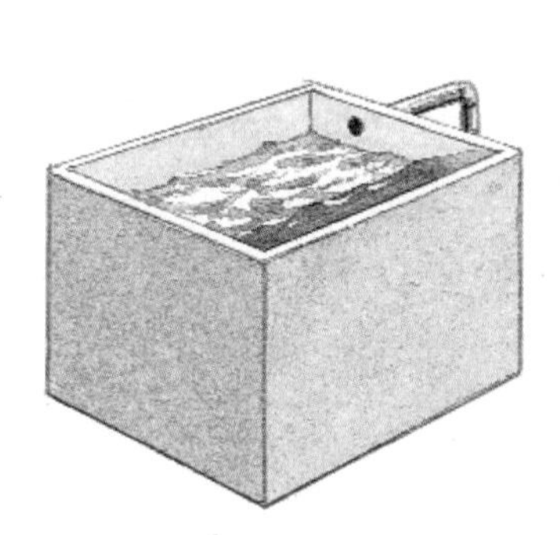

d is the depth in cm; t is the time after midnight in hours.

(a) At midnight, $t = 0$.
What is the depth of the water at midnight?

(b) What is the depth when $t = 2$? What time is that?

(c) What is the depth at 4:30 a.m.?

(d) At what time will the tank be empty?

(e) If you substitute $t = 13$ in the formula, what is the value of d? Does it make sense?

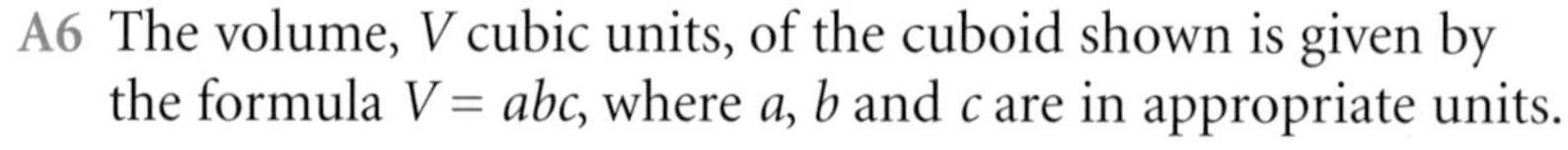

A6 The volume, V cubic units, of the cuboid shown is given by the formula $V = abc$, where a, b and c are in appropriate units.

(a) Show that the surface area, S square units, is given by the formula $S = 2ab + 2bc + 2ca$.

(b) Calculate the volume and the surface area of each of the following. Be careful about the units.

(i) A box 10 cm by 12 cm by 25 cm (ii) A room 8 m by 6 m by 250 cm

A7 A rectangular sheet of card measures 50 cm by 20 cm.
50 of the sheets have a total thickness of 1.0 cm.

(a) Calculate the volume of a single sheet.

(b) Is it sensible to use the formula $S = 2ab + 2bc + 2ca$ to find the suface area of one sheet?

B Further formulas

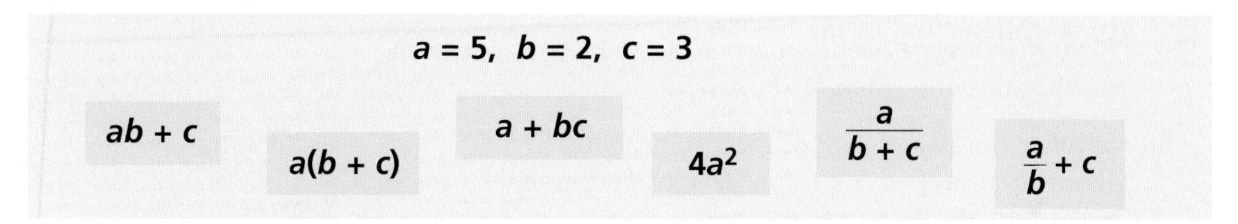

B1 Calculate in your head the values of each of the following when $p = 2$, $q = 3$, $r = 6$.

(a) $r(p + q)$ (b) $rp + q$ (c) $\frac{r}{pq}$ (d) $\frac{q + r}{p}$ (e) $5r^2$

Check that you can make your calculator agree with you.

B2 If $u = 2.1$, $v = 3.1$ and $w = 5.9$, evaluate each of these expressions giving answers accurate to two decimal places.

(a) $\frac{u}{v + w}$ (b) $w - \frac{u}{v}$ (c) $u + vw$ (d) $\frac{u + v}{v + w}$ (e) uv^2

B3 Calculate the values of each of the following in your head when $f = 2$, $g = {}^-6$, $h = 4$.

(a) $f + gh$ (b) $\frac{f}{g - h}$ (c) $(f + g)^2$ (d) $10g^2$ (e) $f^2 + h^2$

Check that you can make your calculator agree with you.

B4 Evaluate each of the expressions in question B3 when $f = 7.9$, $g = 5.7$, $h = 3.8$.

Give your answers accurate to two significant figures.

B5 Evaluate each of these expressions when $a = {}^-4.5$, $b = 0.5$ and $c = {}^-2.5$.

(a) $\frac{a^2}{c^2 - b^2}$ (b) $(c - 2b)^2$ (c) $3b^2$ (d) $(3b)^2$ (e) $ab + bc + ca$

B6 Evaluate each of these expressions when $x = 6$, $y = \frac{1}{3}$ and $z = \frac{1}{2}$.

(a) $xy + z$ (b) $x(y + z)$ (c) $\frac{x}{z} - y$ (d) $\frac{x}{y} - z$ (e) $\frac{x}{z - y}$

B7 The area, A square units, of a trapezium is given by the formula

$$A = \frac{h(a + b)}{2}$$

where h, a and b are the dimensions shown.

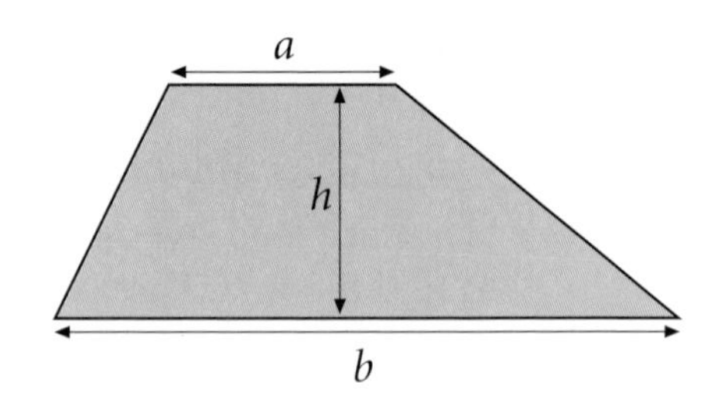

Use the formula to calculate the areas of trapeziums for which

(a) $a = 60\,\text{cm}$, $b = 80\,\text{cm}$, $h = 40\,\text{cm}$

(b) $a = 50\,\text{cm}$, $b = 1.8\,\text{m}$, $h = 80\,\text{cm}$

B8 Calculate the area of each of these trapeziums. Give your answers to a sensible degree of accuracy and include the units.

(a)

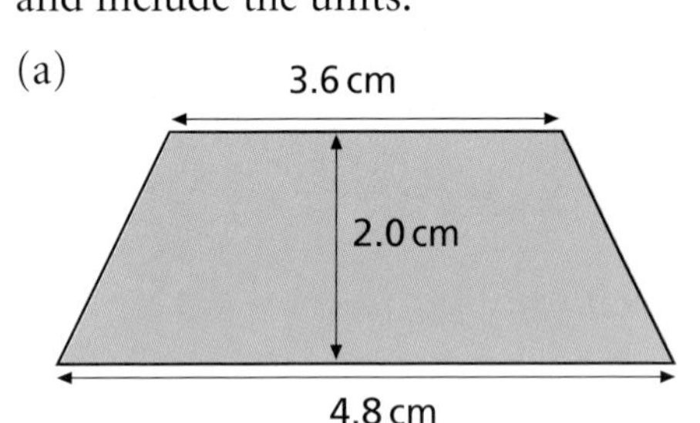

(b)

500 m

600 m

1.2 km

B9 (a) Write down a formula for the volume of the cuboid shown.

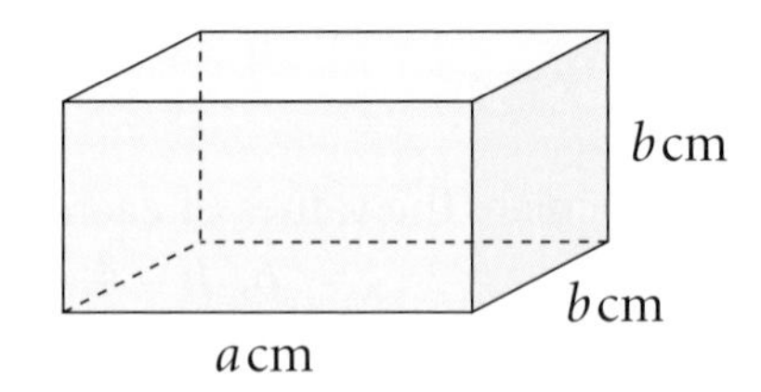

(b) Calculate the volume of a cuboid where $a = 30$ and $b = 40$.

(c) Calculate the volume of a cuboid where $a = 100$ and $b = 100$.

(d) Calculate the volume when

(i) $a = 101$ and $b = 100$ (ii) $a = 100$ and $b = 101$

B10 If $a = 2$ and $b = 3$, calculate in your head the values of

(a) a^2b (b) ab^2 (c) $(ab)^2$

If your results are not all different, you have made a mistake. In that case, find it!

B11 a, b and c are integers between 1 and 5 inclusive. Both $\frac{a}{b+c}$ and $\frac{a}{b-c}$ are also integers.

Find all possible values of a, b and c.

An integer is a positive or negative whole number.

B12 The weight in kg that can be supported at the middle of an oak beam is given by the formula

$$w = \frac{60bd^2}{l}$$

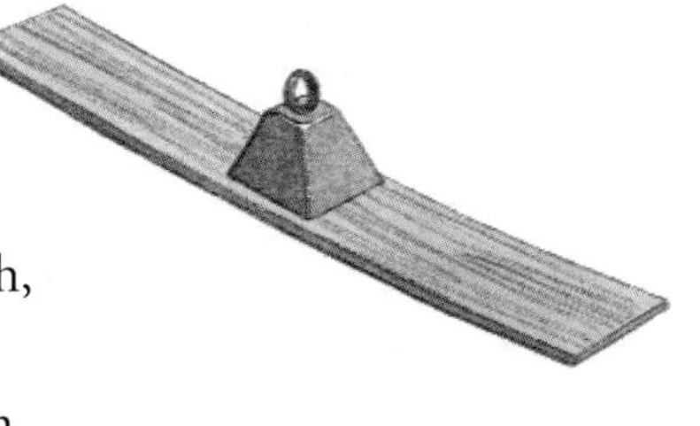

w stands for the weight in kg, and b, d and l for the breadth, depth and length of the beam in cm.

Calculate the load which can be supported by an oak beam 4 m long, 20 cm broad and 30 cm deep.

B13 The dimensions, in cm, of the Sukachu box shown are a, b and c, where $a = 3.5$, $b = 10.0$ and $c = 1.2$.

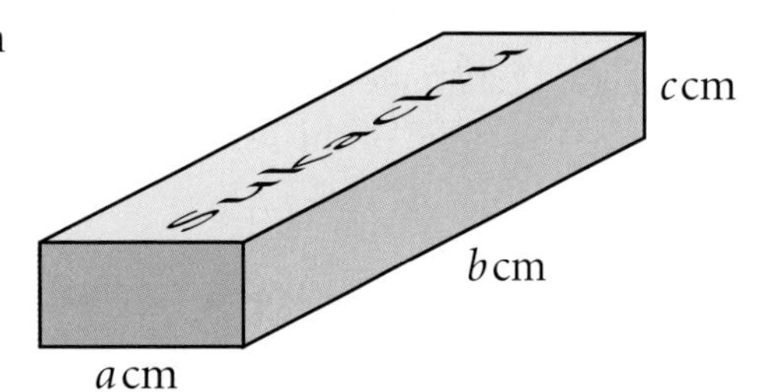

The sales department believe that changes to the dimensions of the box, while keeping the volume (V) the same, would allow a more attractive logo and improve sales.

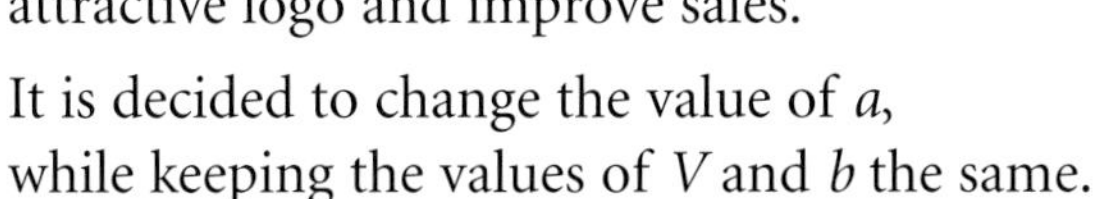

It is decided to change the value of a, while keeping the values of V and b the same.

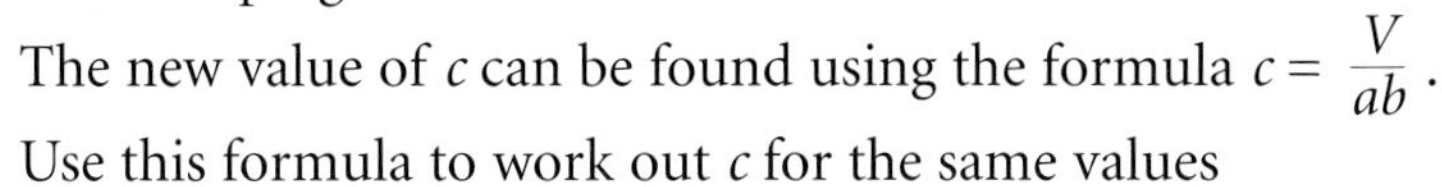

The new value of c can be found using the formula $c = \frac{V}{ab}$.

Use this formula to work out c for the same values of V and b above, but with $a = 4.0$.

B14 A child drops a stone from the top of a cliff which is 80 m above the level of the sea below.

The height of the stone during its fall can be calculated from the formula $h = 80 - 5t^2$.

t is the time in seconds since the stone was dropped.
h is the height of the stone in metres above sea-level.

(a) Calculate h when when t is 0, 1, 2, 3 and 4.
Make a table of the results, like this.

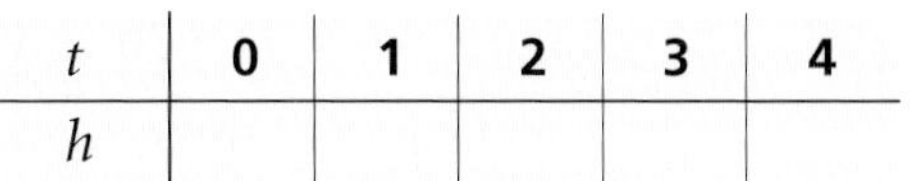

t	0	1	2	3	4
h					

(b) Does the stone fall at the same speed all the time or does it speed up, or does it slow down?

(c) After how many seconds does the stone hit the sea?

B15 Suppose a driver sees a child run out into the road ahead. It takes a little time to think before applying the brakes, and during this time the car travels forward. Then the brakes are applied and the car travels a bit further as it slows down to a stop.

The total distance travelled between the driver seeing the child and the car stopping is called the **stopping distance**. It can be calculated using the formula

$$D = \frac{S(S + 20)}{60}$$

S is the speed of the car, in m.p.h., before the brakes are applied.
D is the stopping distance, in metres.

(a) Calculate D when S is 0, 10, 20, 30, . . . up to 70, and make a table of the results, like this. (Give values of D to one decimal place.)

S (speed in m.p.h.)	**0**	**10**	**20**	**30**	**...**
D (stopping distance in m)					

(b) Draw a graph with S on the horizontal axis, and D on the vertical.

(c) Use your graph to find the speed for which the stopping distance is 50 m.

B16 A factory makes saucepans. The inside surface of each saucepan is coated with a non-stick coating.

When a new size of saucepan is made, the manufacturers need to work out the area of the inside surface, to see how much non-stick coating will be needed.

Here is the formula for working out the area of the inside surface.

$$S = \pi D\left(\frac{D}{4} + d\right)$$

D is the diameter of the saucepan, in cm.
d is the depth of the saucepan, in cm.
S is the inside surface area, in cm^2.

(a) Use the formula to calculate S to three significant figures when $D = 20$ and $d = 13$.

(b) Calculate S to three significant figures when

(i) $D = 24$ and $d = 17$ (ii) $D = 18$ and $d = 10$

(c) What area of non-stick coating will be needed for a pan with a depth of 8 cm and a diameter of 25 cm?

B17 Reference books of drugs used to give only the size of an adult dose of each drug. Nurses had to work out children's doses themselves using 'Young's rule'.

This was $C = \dfrac{An}{n+12}$

C stands for the child's dose, A for the adult's dose, and n for the child's age in years.
(The rule was not used for babies less than 1 year old.)

(a) Use the rule to calculate the dose for an 8-year-old child when the adult dose is 15 milligrams (mg).

(b) A rule used to calculate the dose for a child less than 1 year old was called 'Fried's rule'.

The rule was $C = \dfrac{Am}{150}$, where m is the child's age in months.

Calculate the dose for a 9-month-old child when the adult dose is 400 mg.

***B18** A grandfather clock has a pendulum with an adjustable length.
The time for one complete swing from left to right and back again is given by the formula $T = 2\pi\sqrt{\dfrac{L}{g}}$.
T is the time for a complete swing in seconds;
L is the length of the pendulum in metres;
g is the gravitational acceleration (in ms^{-2}) wherever the clock is situated.

(a) In London, $g = 9.807$. What is the time for one complete swing, to three decimal places, if $L = 0.990$?

(b) On the equator, $g = 9.780$. At the North Pole, $g = 9.832$.
If $L = 0.990$, how many complete swings will the pendulum make in one hour

(i) on the equator (ii) at the North Pole

(c) The formula for L when you know T is $L = \dfrac{gT^2}{4\pi^2}$

How long should the pendulum be so that one complete swing at the North Pole is, as nearly as possible, 2.00 seconds?

What progress have you made?

Statement	Evidence
I can interpret simple formulas.	**1** The depth of water in a reservoir is given by $d = 2 + 0.2t$ d is the depth of the water in metres; t is the time after noon in hours. (a) When $t = 2$, it is 2 p.m. What is the depth of water then? (b) What is the depth of water at 6 p.m.? (c) What is d at 7:30 p.m.? (d) What is the depth of water 30 minutes after noon?
I can substitute in formulas involving letters.	**2** Work out, without using a calculator, the value of each of these when $a = 2$, $b = 3$ and $c = 4$. (a) $a(b + c)$ (b) $\frac{a + c}{b}$ (c) $\frac{c}{b - a}$ (d) $a + bc$ (e) ab^2 (f) $a^2 + c^2$
	3 Evaluate each of these expressions when $p = 0.6$, $q = ^-1.8$ and $r = ^-1.2$. (a) $pq - r$ (b) $\frac{p - q}{r}$ (c) $p - \frac{q}{r}$ (d) pq^2
I can substitute values into more complex formulas.	**4** It can be shown that if an object is thrown upwards with speed u m/s, then after a time t seconds its height will be h m, where $h = \frac{t(2u - 9.8t)}{2}$ (a) Find h (to the nearest whole number) when $u = 62.5$ and $t = 4.5$. (b) A stone is thrown upwards with speed 88 m/s. How high will it be 15 seconds later?

24 Locus

This work will help you

- understand what is meant by a locus
- draw loci

A Place the points

This activity is described in the teacher's guide.

B Regions and boundaries

You need sheet 208.

The map shows an island, with houses shown as crosses (+).
The island is subject to very bad weather at times, so a warning system is needed.

B1 One idea is to build a radio transmitter at A, with a range of 5 kilometres.

(a) Mark all the houses that will be within range of the transmitter.

(b) What is the boundary of the region containing all points within range?

B2 A better idea is to increase the range of the transmitter to 7 km.
Draw the boundary of the region which this transmitter would cover.

B3 Because many houses would still be out of range, it is suggested that two 5 km transmitters could be used.

Mark where you would put them so that as many houses as possible are in range.

B4 In an extreme emergency it may be necessary to evacuate the island.
Two evacuation stations are built on the coast at B and C.
People living in the houses go to the nearest station.

Mark clearly

(a) all the houses which are nearer to B than to C

(b) those which are equidistant from B and C

(c) those which are nearer to C than to B

(d) the boundary separating the regions covered by stations B and C

C Sets of points

The word **locus** means the set of all the points that fit a given description.

In this diagram, the point A is fixed.

The locus of points that are 3 cm from A is the circle with centre A and radius 3 cm.

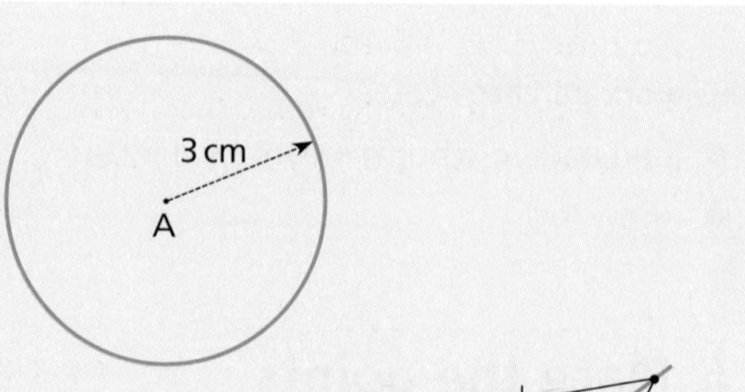

The locus of points that are equidistant from two points B and C is the perpendicular bisector of BC.

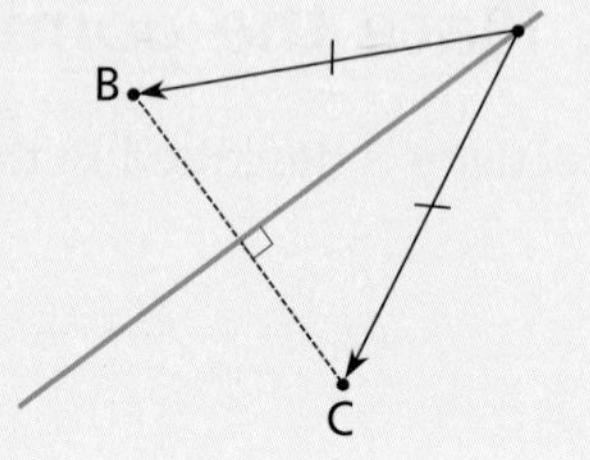

C1 A field is in the shape of an equilateral triangle with sides of length 100 m.
There is an unexploded bomb at one corner of the field.
It is dangerous to be within 50 m of the bomb.

Draw a diagram and shade the part of the field that is safe.

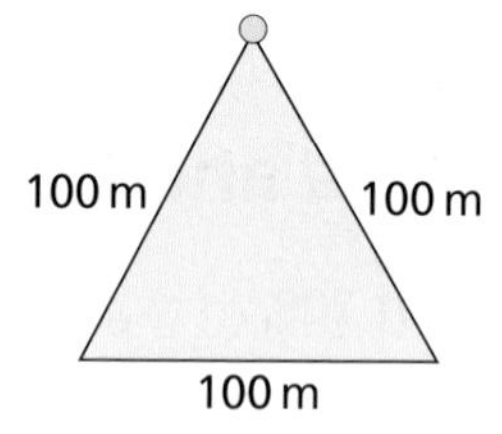

C2 Repeat C1, but this time there is an unexploded bomb at every corner of the field.

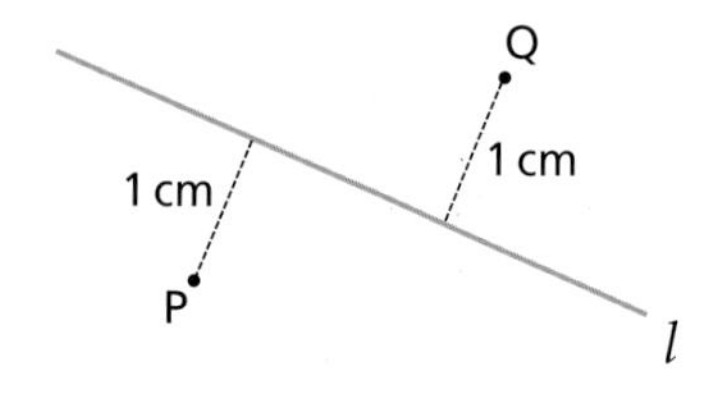

C3 The line l is fixed.
P and Q are examples of points that are 1 cm from l.

Draw a line l and draw the locus of all the points that are 1 cm from l.

C4 The diagram shows a rectangular yard, 6 m by 4 m.
A power line stretches across the yard as shown.

It is dangerous to stand closer than 1 m to the power line.
Draw the rectangle to scale and shade the region of the yard where it is dangerous to stand.

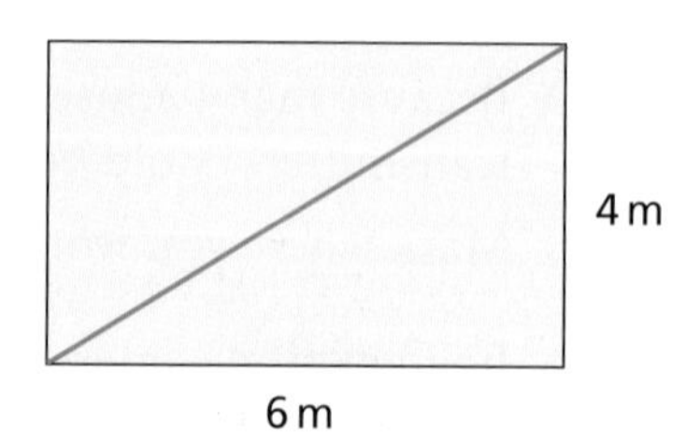

3 m T
8 m
4 m
10 m

C5 Dawn's garden is a rectangle 10 m by 8 m.
The point T shows the position of a tree.

Dawn wants to plant another tree in the garden.
It must be at least 2 m away from the edges of the garden and at least 3 m from T.

Draw a diagram and shade the region where the new tree could be planted.

C6 ABCD is a square.
P is an example of a point inside the square that is equidistant from the sides AB and AD.

Draw the square and the locus of all the points that are inside the square and equidistant from AB and AD.

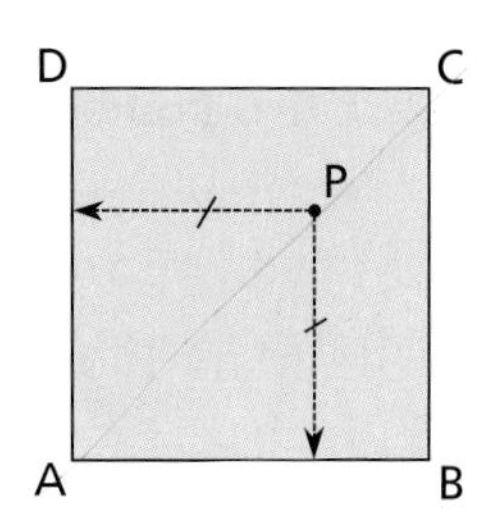

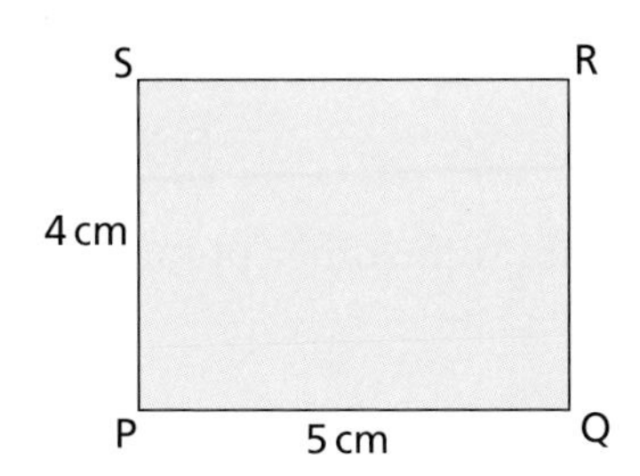

C7 PQRS is a rectangle. PQ = 5 cm, PS = 4 cm.

Draw the rectangle and draw the locus of all the points inside the rectangle that are equidistant from the sides PQ and PS.

C8 The lines l and m are fixed.
P is an example of a point that is equidistant from l and m.

Draw l and m and the locus of all points equidistant from l and m.

Describe the locus in words.

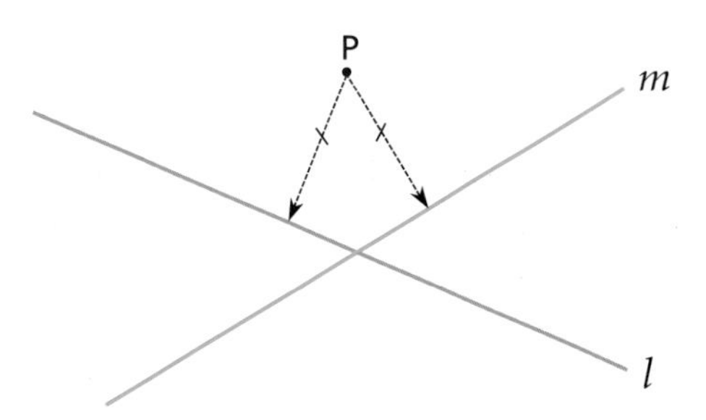

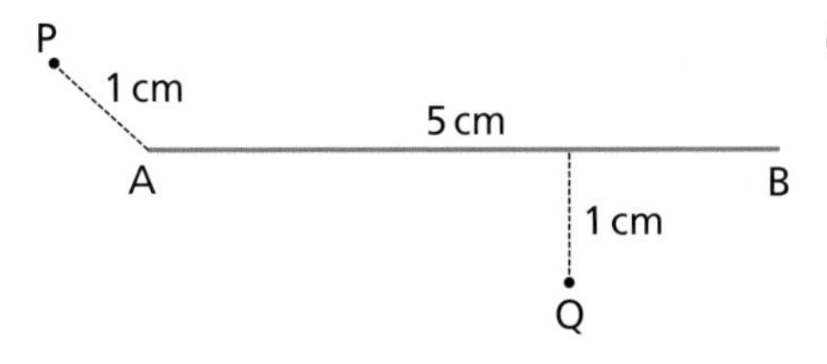

C9 AB is a line segment of length 5 cm.
P and Q are examples of points that are 1 cm from the nearest point of AB.

Draw AB and the locus of all points that are 1 cm from the nearest point of AB.

C10 The diagram shows part of the boundary between two countries.
The governments agree that there should be a 'no go' zone on each side of the boundary.

The zone is to consist of all points within 1 km of the boundary.

Draw a diagram and shade the 'no go' zone.

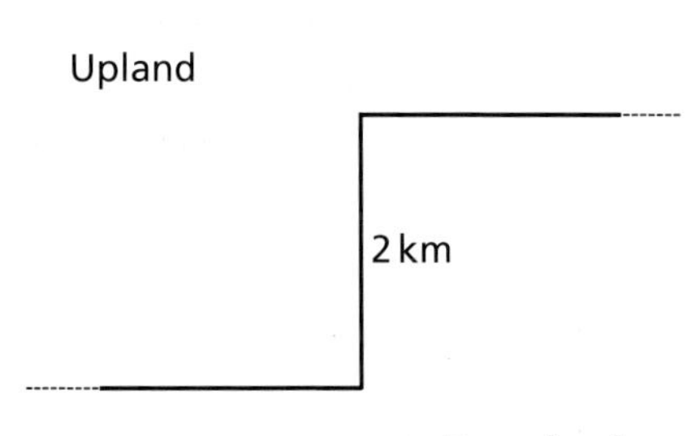

C11 Mark two points A and B, 8 cm apart.

- Draw an arc, centre A, with radius 6 cm, and an arc, centre B, with radius 4 cm.
- Mark P, one of the points where these arcs cross.

The total distance PA + PB is 10 cm.
So P is one of the points on the locus of points whose total distance from A and B is 10 cm.

(The other point where the arcs cross is also on the locus.)

- By drawing other arcs whose radii add up to 10 cm, find some other points of this locus. Then draw the locus, as accurately as you can.

(Another way to draw the locus is to put two pins at A and B and use a loop of thread of total length 18 cm, as shown here.)

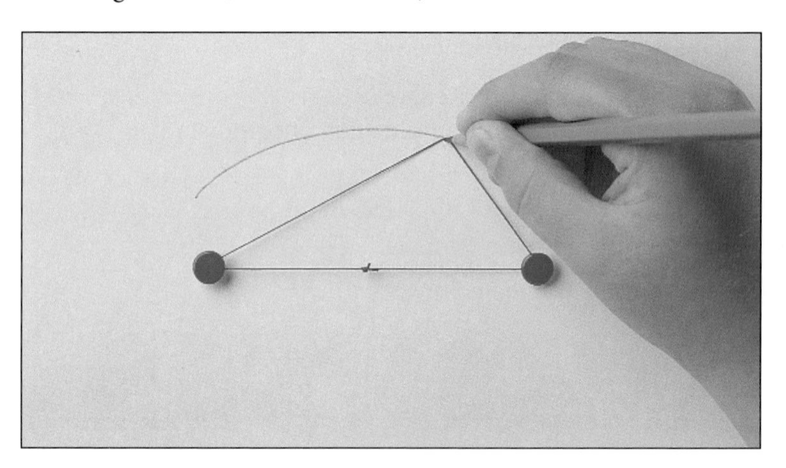

***C12** Draw a line l and a point A 4 cm away from it.

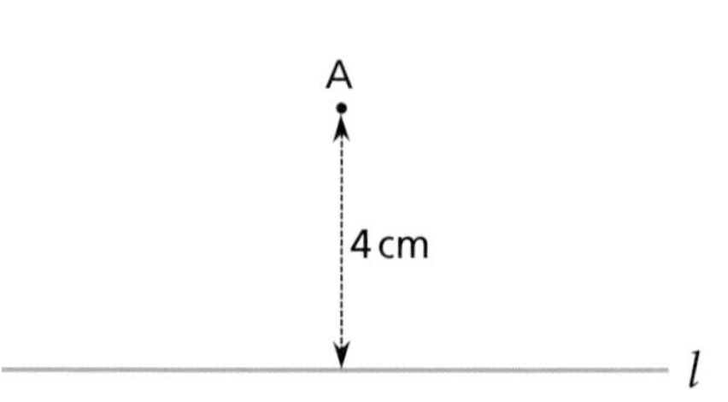

- Draw an arc, centre A, radius 3 cm.
- Draw a line parallel to l and 3 cm from l.

The points P and Q where the arc and the line cross are both on the locus of points equidistant from A and l.

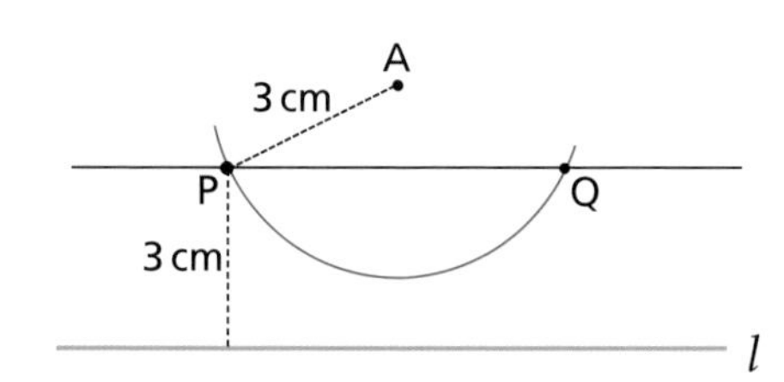

- By drawing other arcs and lines, find some more points on this locus, and draw the locus.

D Moving points

The word 'locus' is also used to describe the path of a point that moves according to a rule.

D1 AB is a ladder, 5 m long. M is the midpoint of AB. As the ladder slides down the wall, the point M traces out a locus.

Make a scale drawing of the ladder in different positions as it slides down the wall. (2 cm to 1 m is a suitable scale.) Draw the locus of M.

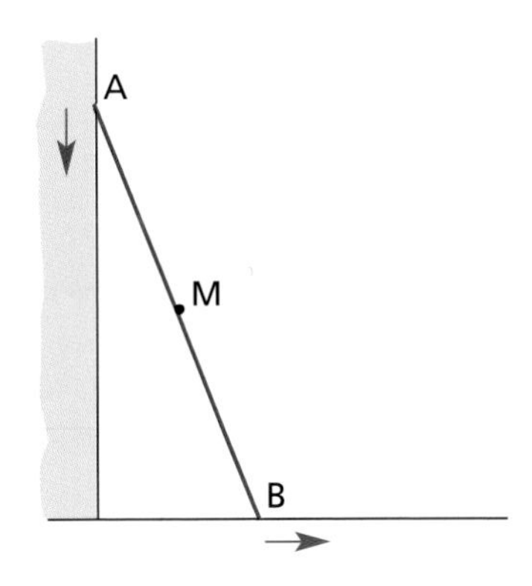

D2 Repeat the previous question, but this time find the locus of the point P on the ladder, where AP = 2 m.

D3 The end of a box is a rectangle ABCD, 60 cm by 40 cm. The box is on flat ground and is rolled over and over as shown.

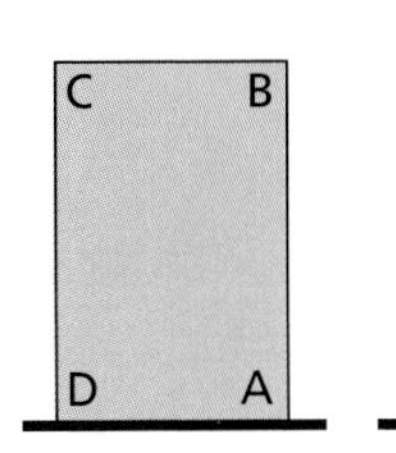

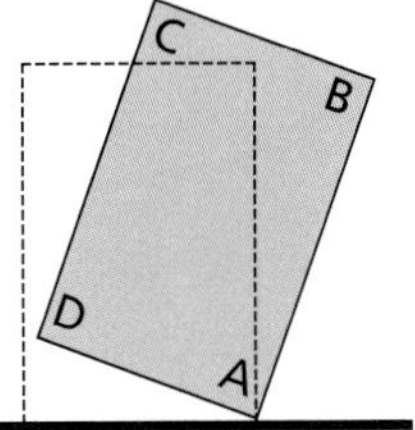

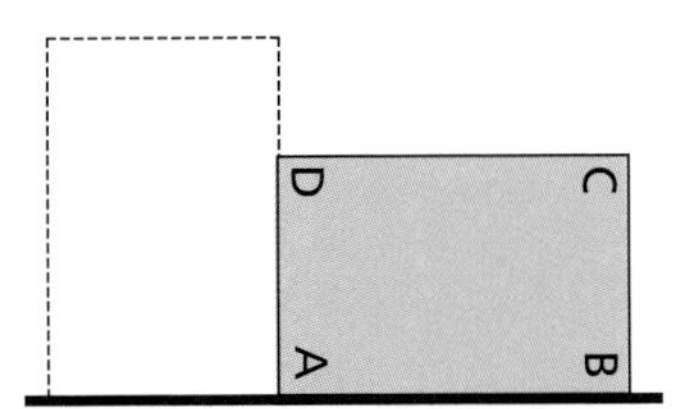

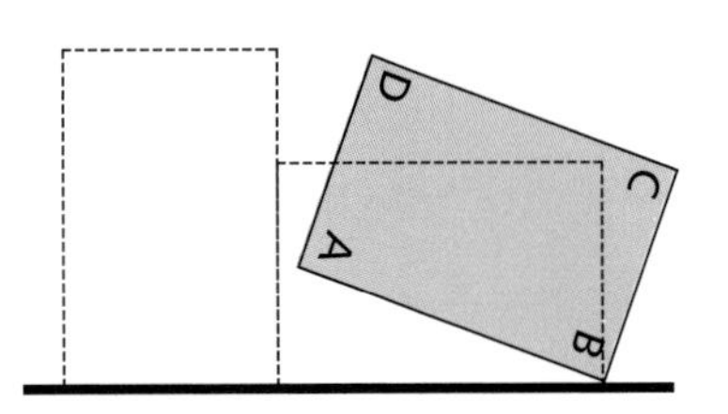

The box is rolled until it is upright again with A and D on the ground.
Draw to scale the five positions of the rectangle on the ground.
Draw the locus of the point B as the box rolls.

D4 A rod rotates about one end at a constant rate of 10° per second.
As it does so, an insect walks along the rod at a speed of 0.5 cm per second, starting at the fixed end.

Draw the locus of the insect as it moves.

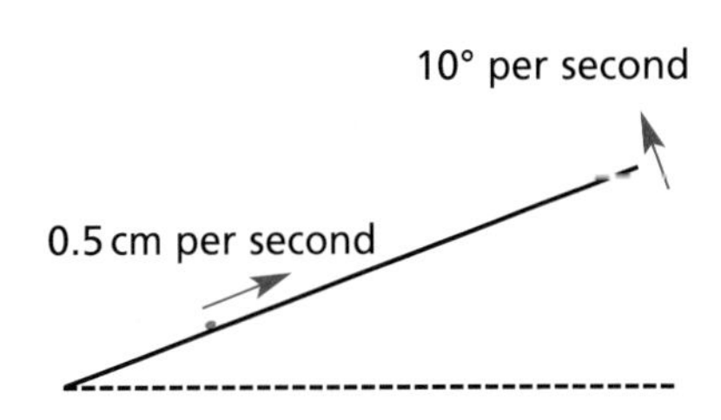

***D5** This diagram shows part of the mechanism of a funfair ride.

A is fixed. The bars AB and BC are jointed at B.
To start with, A, B and C are in a straight line.

The bar AB rotates about A at 10° per second.
The bar BC rotates about B at 20° per second.

Draw the locus of C.

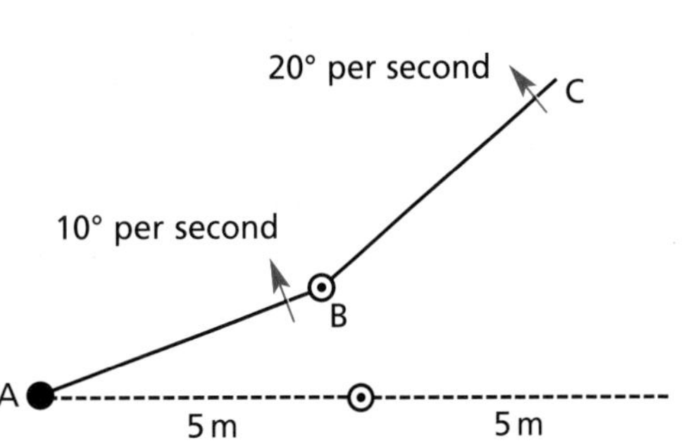

*E Loci in three dimensions

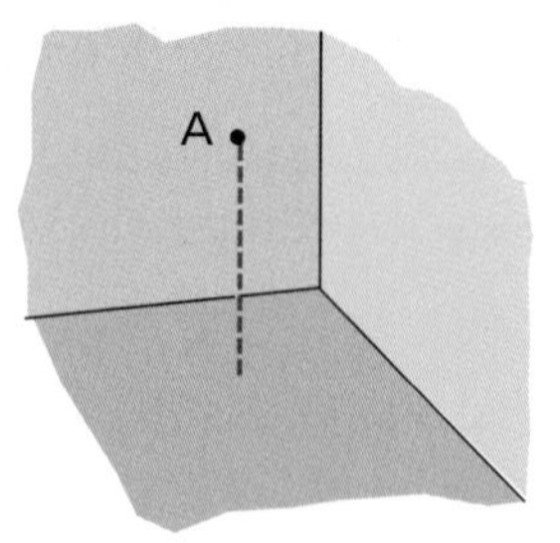

E1 A point A is fixed in three-dimensional space.
Describe the locus of all points that are 1 metre from A.

E2 Two points A, B are fixed in three-dimensional space.
Describe the locus of all points that are equidistant from A and B.

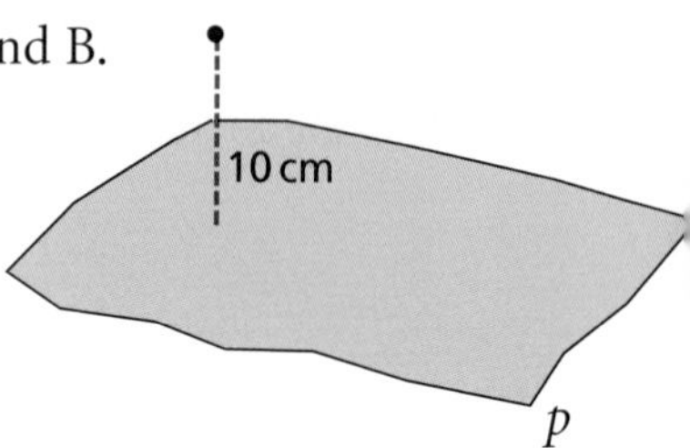

E3 The plane *p* is fixed in three-dimensional space.
Describe the locus of all points that are 10 cm from *p*.

E4 A line *l* is fixed in three-dimensional space.
Describe the locus of all points that are 10 cm from *l*.

E5 The line *l* is fixed in three-dimensional space.
A is a fixed point on *l*.

AP is a line segment of fixed length.
It moves so that the angle between AP and the line *l* is always 30°.

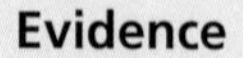

Describe

(a) the locus of the point P

(b) the three-dimensional shape traced out by the segment AP

What progress have you made?

Statement	Evidence
I can draw the locus of points that fit a given description.	**1** ABCD is a rectangle 4 cm by 3 cm. Draw the locus of points inside the rectangle that are equidistant from A and C.
I can draw the locus of a point that moves according to a given rule.	**2** This rectangle is rotated about the midpoint of the side AB. Draw the locus of C.

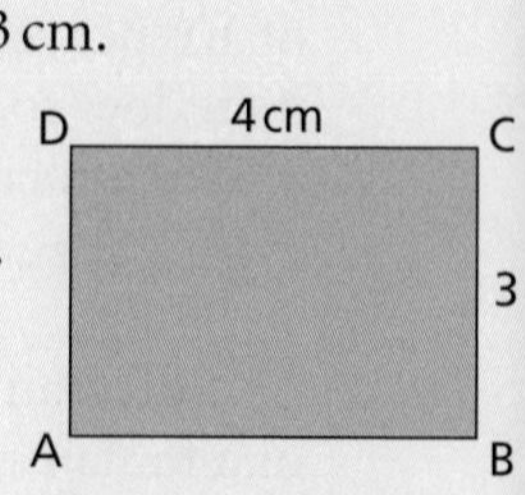

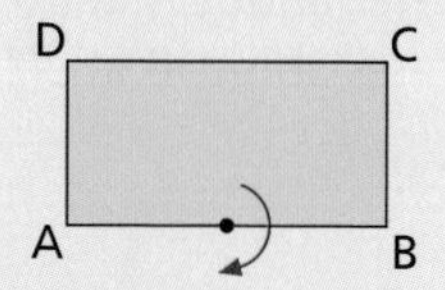

25 Distributions

This work will help you

- interpret and draw stem-and-leaf tables and frequency polygons
- revise mode, median, mean and range
- calculate or estimate the mean of a frequency distribution

A Median, range and mode

A1 Here are the weights in kg of a group of children.

(a) Find the median weight.

(b) Find the range of the weights.

37 51 41 45 52
57 39 42 56 46
46 50 48 38 41

A2 Find the median and range of these heights in cm.

139 151 128 126 147
129 139 140 142 156
160 145

A3 Charlie watched cars go by his house and noted down how many people were in each car. Here is his data, written out in order:

1, 1, 1, 1, 1, 1, 2, 2, 3, 3, 3, 3, 3, 3, 3, 4, 4, 4, 5

What is the median number of people in a car?

A4 Paula also noted the number of people in cars. She made this tally table.

(a) How many cars did Paula observe?

(b) What is the median number of people in a car?

Number in car		Frequency
1	𝍸 𝍸 𝍸	15
2	𝍸 ////	9
3	𝍸 //	7
4	𝍸 /	5
5	///	3

A5 Zak summarised his car data in this frequency table.

Find the median number of people in a car.

Number in car	1	2	3	4	5
Frequency	7	10	11	6	4

A6 The most frequently occurring number of people is called the **mode.** What is the mode for each of the data sets in A3, A4 and A5?

A7 The heights in cm of the players in a seven-a-side football team are

163 169 165 165 168 171 161

All of these heights are over 160 cm.
The 'extra' heights above 160 cm are

3 9 5 5 8 11 1

(a) Work out the mean of these extra heights.

(b) Write down the mean height of the seven team members.

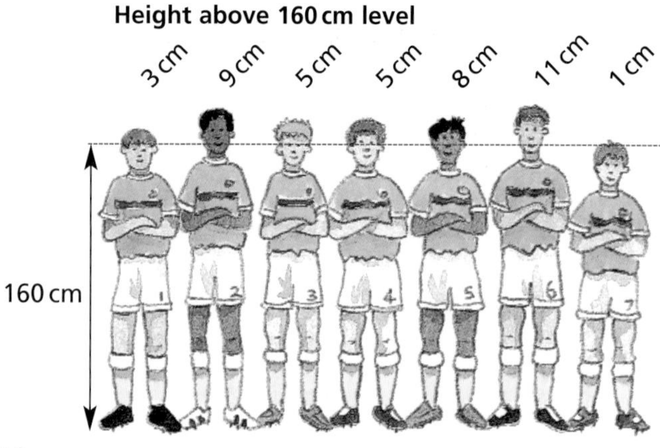

A8 Perry counted the matches in each of 12 boxes.

The numbers were 38, 35, 33, 37, 36, 35, 39, 40, 36, 32, 30, 38

Use a method similar to that in A7 to work out the mean number of matches.

A9 The ages of the players in a seven-a-side team are

10 years 6 months 10 years 4 months 11 years 2 months 10 years 4 months
11 years 6 months 10 years 1 month 10 years 9 months

(a) Work out the mean age of the team in years and months.

(b) What is the range of the ages?

A10 Sharmila drove from London to Birmingham six times last month.
Her journey times were

2 hours 35 minutes 2 hours 50 minutes 3 hours 10 minutes
2 hours 55 minutes 3 hours 15 minutes 2 hours 45 minutes

Work out the mean journey time.

A11 The mean weight of the 11 players in a football team is 72 kg.

(a) What is the total weight of the team?

(b) A player who weighs 68 kg is replaced by a player weighing 73.5 kg.
What is the mean weight of the team afterwards?

A12 The mean weight of the 15 players in a rugby team was 87 kg.
Then Joe, who weighed 90 kg, was replaced by Karl.
Afterwards the mean weight was 86 kg.

How much did Karl weigh?

A13 There were 11 people living in a retirement home.
Then Peggy died at the age of 94.
The mean age of the remaining 10 people was 72.

What was the mean age before Peggy's death?

B Stem-and-leaf tables

An introductory activity is described in the teacher's guide.

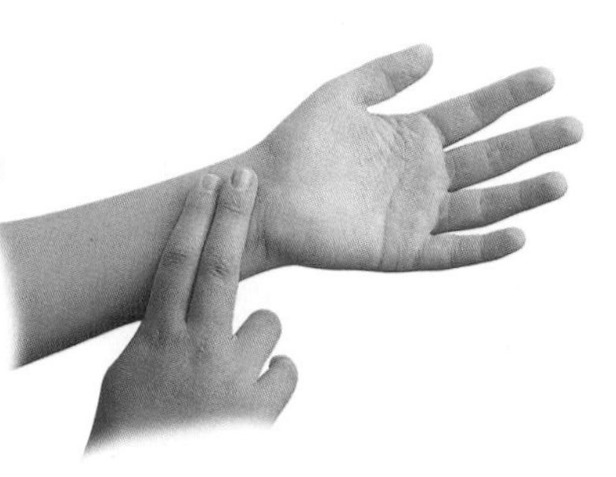

B1 This stem-and-leaf table shows the marks of some pupils in a test. Find

(a) the range of the marks

(b) the number of pupils who took the test

(c) the median mark

(d) the modal group

Stem	Leaves
4	3 8
5	0 4 4 7
6	2 4 7 9 9
7	1 3 3 5 5 8 8 9
8	0 0 4 5 6
9	0 1 1

B2 Thirty pupils took a maths exam. There were two papers, each marked out of 100. Here are the results.

Paper 1	58	43	82	66	70	41	49	28	62	54
	80	44	72	70	45	49	55	61	54	63
	39	60	54	79	66	50	38	63	66	82
Paper 2	48	44	63	54	69	70	40	37	61	58
	56	35	42	51	28	34	39	41	44	38
	62	49	52	29	38	83	47	50	64	55

Make a stem-and-leaf table for each paper.
Which paper appeared to be harder? How can you tell?

B3 These are the weights in kg of 20 newborn babies.

3.0	3.2	4.1	3.8	3.7	2.5	1.7	2.3	3.3	2.8
1.9	3.1	2.6	3.3	2.5	4.0	2.7	3.4	2.8	1.9

(a) Make a stem-and-leaf table with the whole numbers as stems and the tenths as leaves. The table has been started here.

Stem	Leaves
1.	
2.	
3.	0
4.	

(b) What is the range of the weights?

(c) What is the median weight?

C Mean

Here are the numbers of people in 20 cars which passed a ticket booth.

1 1 1 1 1 1 2 2 2 2 3 3 3 3 3 3 3 4 4 5

We calculate the mean number of people per car by dividing the total number of people by the number of cars.

Total number of people = 1 + 1 + 1 + 1 + ... + 4 + 5 = 48

Mean number of people per car = $\frac{48}{20}$ = **2.4**

C1 In the list above, 1 occurs 6 times, 2 occurs 4 times, and so on.

This is shown in this frequency table.

Number in car	Frequency
1	6
2	4
3	7
4	2
5	1

Show how to find **from the frequency table**

(a) the total number of cars

(b) the number of cars which had 3 people in them

(c) the total number of people who were in cars with 3 people in them

(d) the total number of people in all the cars

(e) the mean number of people per car

C2 This frequency table comes from a survey of the houses in a street.

Number in house	Frequency
0	2
1	3
2	6
3	5
4	2
5	3

(a) How many houses have 2 people living in them?

(b) How many houses are there in the street?

(c) How many people are there altogether in the houses which have 3 people in them?

(d) How many people are there altogether in the street?

(e) What is the mean number of people per house to 1 d.p.?

The calculation of the mean from a frequency table is often set out by putting an extra column in the table.

Number in car	Frequency	Number × frequency
1	6	6
2	4	8
3	7	21
4	2	8
5	1	5
Totals	20 cars	48 people

7 cars with 3 people each

$$\text{Mean} = \frac{48}{20} = \mathbf{2.4}$$

C3 This data comes from a survey of birds' nests.

Calculate

(a) the number of nests surveyed

(b) the total number of eggs in all the nests

(c) the mean number of eggs per nest, correct to 1 d.p.

Number of eggs in nest	Frequency
3	16
4	14
5	4

C4 Paula opened some boxes of paper clips and counted how many there were in each box. Her results are shown in this frequency table.

Calculate

(a) the number of boxes Paula opened

(b) the total number of paper clips in all the boxes

(c) the mean number of clips per box, correct to 1 d.p.

Number of paper clips in box	Frequency
32	4
33	7
34	9
35	20
36	10

C5 Hitesh counted the number of flowers on each stem of a species of plant. His results are shown in this bar chart.

Calculate

(a) the number of stems Hitesh looked at

(b) the total number of flowers on all of the stems

(c) the mean number of flowers per stem, correct to 1 d.p.

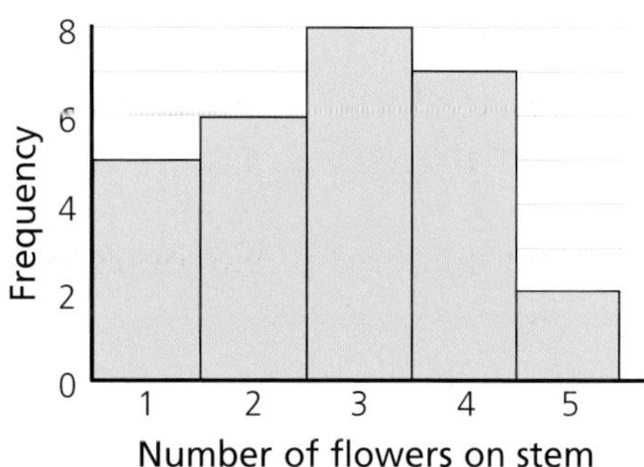

D Discrete and continuous data

Discrete data usually comes from counting.

For example, the number of tomatoes on a plant can be 0, 1, 2, 3, …
but not numbers in between, like 1.7, 2.08, and so on.

This data about tomatoes has been **grouped**: 0–4, 5–9, 10–14, …

Number of tomatoes	0–4	5–9	10–14
Number of plants	3	6	8

Because the data is discrete, there are jumps between the groups.

Continuous data usually comes from measuring.

If a baby weighs between 2 kg and 3 kg, its weight could be anything between 2 kg and 3 kg, for example 2.473 kg or 2.856 235 41 kg.

It depends on the accuracy of the weighing machine!

This dot plot shows the birth weights, in kg, of 16 babies.

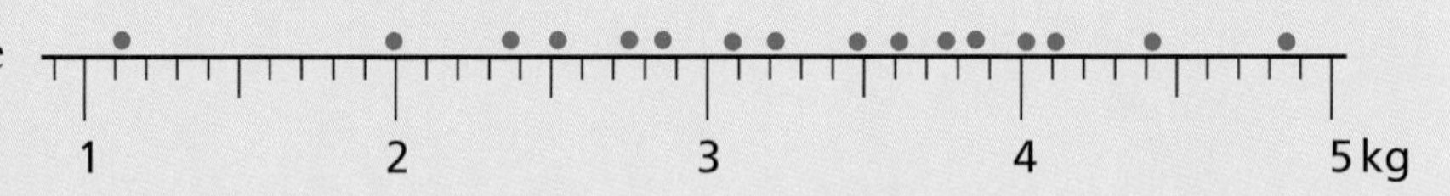

We can group the weights into **intervals**, for example 1–2 kg, 2–3 kg, 3–4 kg, …
There are no jumps between the intervals, because the data is continuous.
But we have to decide where to put a weight of exactly 2 kg or 3 kg, and so on.

In this table and chart, the 2 kg baby has been included in the **upper** interval 2–3 kg.

Weight in kg	1–2	2–3	3–4	4–5
Frequency	1	5	6	4

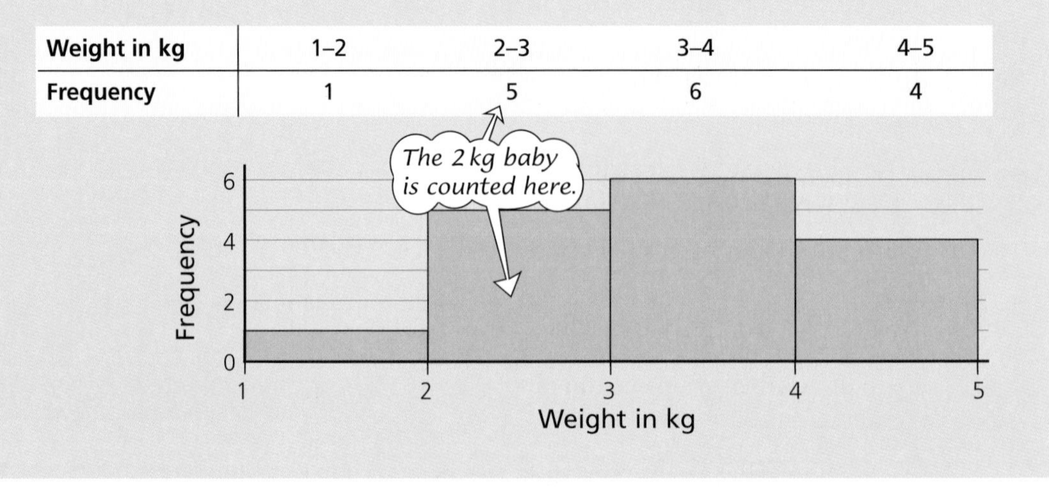

D1 (a) Use the dot plot to find the range of the babies' weights, as accurately as you can.

(b) Why is it not possible to find the range from the frequency chart?

D2 Is each of these a discrete quantity or a continuous quantity?

(a) Time taken to run 100 m
(b) Average speed of traffic
(c) Age (given as the number of whole years)
(d) Shoe size

D3 The teachers in a primary school asked all the children how long it took them to get home from school.
This chart shows the frequency distribution of the journey times.

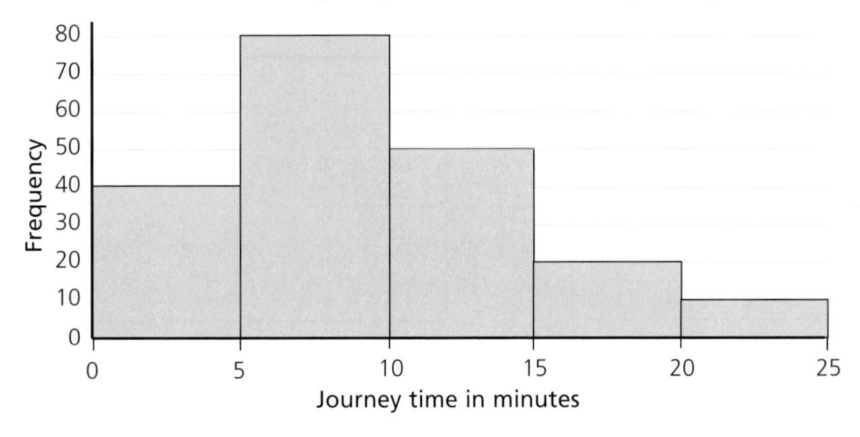

(a) Karl says that the longest journey time is 25 minutes.
Is he right? Explain your answer.

(b) Why is it not possible to find the range of the journey times from the chart?

(c) Which is the modal interval?

(d) What percentage of the children have journey times in the modal interval?

(e) If school ends at 3:45 p.m., what percentage of the children would get home at 4 p.m. or later?

D4 Draw a rough sketch of the kind of frequency chart of journey times you would expect to get for each of these schools.

(a) A school in a housing estate where all the pupils live very close to the school.

(b) A school which serves two villages some distance apart.
The school is situated in one of the villages.
About half the pupils come from this village and about half from the other.

(c) A school where most of the pupils live nearby, but a few live quite a long way away.

(d) A school where all the pupils live a long way from the school.

E Frequency polygons

A **frequency polygon** is another type of frequency graph.

Instead of drawing bars …

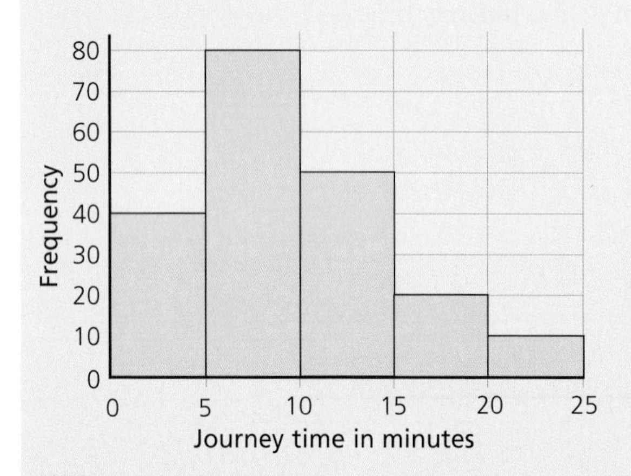

… plot a point at the **mid-point** of each interval and join them up with straight lines.

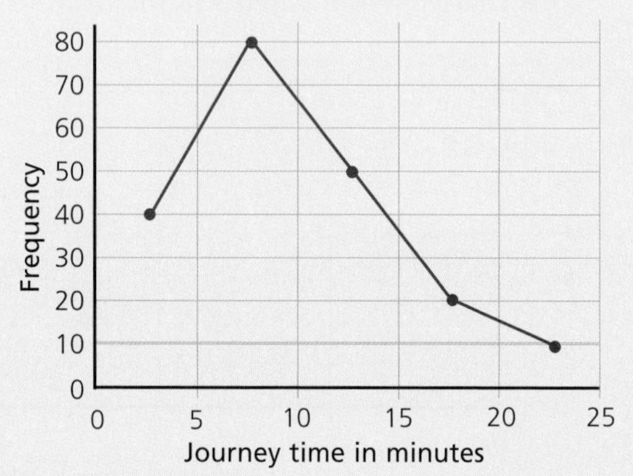

One advantage of frequency polygons is that it is easier to show two distributions on the same graph, for comparison.

E1 These are the frequency polygons of journey times to home from two schools A and B.

(a) How many pupils take between 5 and 10 minutes to get home from school B?

(b) Which is the modal interval for school A?

(c) Which is the modal interval for school B?

(d) In a few words, compare journey times for the two schools.

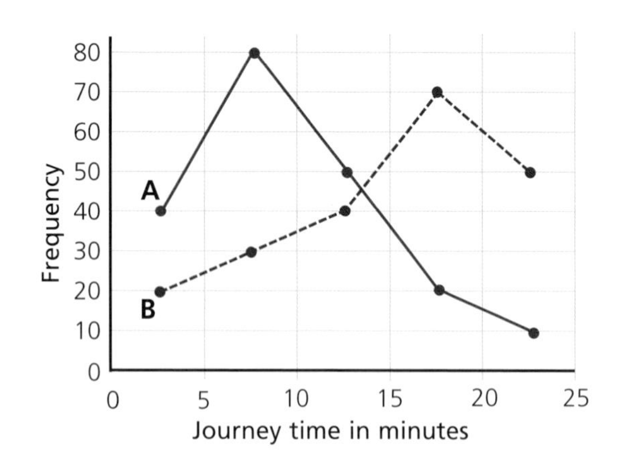

E2 This table gives information about some boys' and girls' birth weights. Draw two frequency polygons on the same axes.

Weight in kg	0.0–1.0	1.0–2.0	2.0–3.0	3.0–4.0	4.0–5.0
Number of boys	2	4	9	6	3
Number of girls	1	9	7	5	2

F Mean of a grouped frequency distribution

Here is the frequency table for the heights of some children.

Height in cm	Frequency
120–130	4
130–140	7
140–150	6
150–160	8

We can't calculate the mean height because we don't know the actual heights.

But we can estimate the mean by using **mid-interval values.**
The mid-interval value for the interval 120–130 cm is halfway between 120 and 130. This is 125 cm.

We treat each of the 4 children in this interval as if their height was 125 cm.

The calculation of the estimate is often set out like this:

Height in cm	Mid-interval value	Frequency	Mid-interval value × frequency
120–130	125	4	500
130–140	135	7	945
140–150	145	6	870
150–160	155	8	1240
	Totals	25 children	3555 cm

4 children each of height 125 cm

$$\text{Estimated mean height} = \frac{\text{total height}}{\text{number of children}} = \frac{3555}{25} = \mathbf{142.2\,cm}$$

F1 This frequency table summarises information about the weights of a group of children.

(a) Copy and complete the table.

(b) Calculate an estimate of the mean weight of the group.

Weight in kg	Mid-interval value	Frequency	Mid-interval value × frequency
20–30	25	6	
30–40		5	
40–50		3	
50–60		2	
	Totals	.... children	...... kg

F2 Calculate an estimate of the mean weight of the people whose weights are summarised in this table.

Weight in kg	Frequency
40–45	3
45–50	6
50–55	10
55–60	4

F3 Calculate an estimate of the mean weight of the apples whose weights are summarised in this table.

Weight in g	70–85	85–100	100–115	115–130
Frequency	6	10	8	5

F4 This tally chart shows the distribution of the ages of the members of a club.

(a) Calculate an estimate of the mean age of the members.

(b) In which age group does the median age lie?

Age group		Frequency
16–20	~~////~~ /	6
21–25	~~////~~ ~~////~~	10
26–30	~~////~~ ~~////~~ //	12
31–35	~~////~~ ~~////~~ ~~////~~	15
36–40	////	4

F5 This frequency polygon shows the distribution of journey times home from a school in the country.

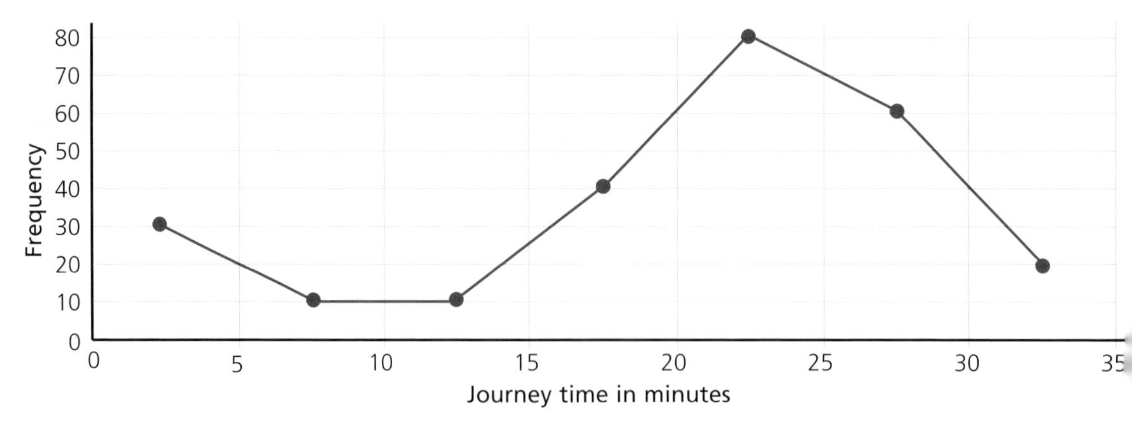

(a) What can you say about the longest journey time?

(b) What is the modal interval?

(c) What percentage of the children at this school take 20 minutes or more to get home?

(d) If school finishes at 3:45 p.m., what percentage of the children would get home at 4 p.m. or later?

(e) If school were to finish at 3:35 p.m., what percentage of the children would get home at 4 p.m. or later?

(f) Calculate an estimate of the mean journey time.

G Summarising and comparing data

It is often useful to summarise a set of data. This may make it easier to compare it with other data.

To summarise data we need

some idea of the 'average' of the data

some idea of how spread out the data is

Measures of average

The word 'average' does not have a precise meaning. It is used to mean 'typical', 'representative', 'round about the middle, neither big nor small'.

In statistics, which is the science of handling data, there are several ways of getting a measure of average. Three of these are the **mode**, **median** or **mean**.

Often one of these is a better measure of 'average' than the others.
It sometimes happens that none of them is a good 'average'.

G1 Here are the salaries of the people who work for a small company. ('k' means thousand.)

£8k, £9k, £9k, £9k, £11k, £11k, £13k, £30k, £53k

(a) Calculate the mean salary.
Does it give a good idea of a typical salary in the company?

(b) Write down the median salary and the modal salary.
Do either of these give a better idea of a typical salary?

G2 In another company the salaries are

£12k, £12k, £12k, £13k, £13k, £14k, £14k, £30k, £31k, £31k, £32k, £32k, £33k

(a) Find the mean, median and mode.

(b) Which gives the best idea of a typical salary in this company?

Measures of spread

Several different measures of spread are used in statistics.
The one we have met so far is the simplest: **range**.

G3 These are the salaries in two different companies.

Company A: £12k, £8k, £28k, £19k, £14k, £18k, £8k, £24k, £13k

Company B: £17k, £15k, £12k, £22k, £11k, £16k, £12k, £23k

Calculate the mean and range of each company's salaries.
Write a couple of sentences comparing the two companies.

*G4 There are 9 employees in a company.
The lowest salary is £10k and the mean salary is £20k.

For each of these statements say whether it must be true, may be true or must be false. Explain your choices.

(a) The total of all the salaries is £180k.
(b) The median salary is £20k.
(c) The median salary is £10k.
(d) The range of the salaries is £110k.

What progress have you made?

Statement / **Evidence**

I can use a stem-and-leaf table.

1 This stem-and-leaf table shows the ages of the people in a choir.

2	4 5 5 7
3	0 2 5 6 6 8 8
4	2 3 4 4 6 8 9 9
5	0 2 3 3
6	2 7

(a) How many people are there in the choir?
(b) What is the median age?
(c) What is the range of the ages?

I can draw a frequency polygon.

2 Draw a frequency polygon to show this data about the weights of the pupils in a class.

Weight in kg	Frequency
30–35	3
35–40	6
40–45	10
45–50	4
50–55	2

I can calculate an estimate of the mean of grouped data.

3 Calculate an estimate of the mean weight of the pupils in the table above.

Review 3

1 Houseproud DIY sells rectangular pieces of silvered glass for mirrors.
The cost of a piece is proportional to its area.

Piece A costs £9. Work out the costs of pieces B and C.

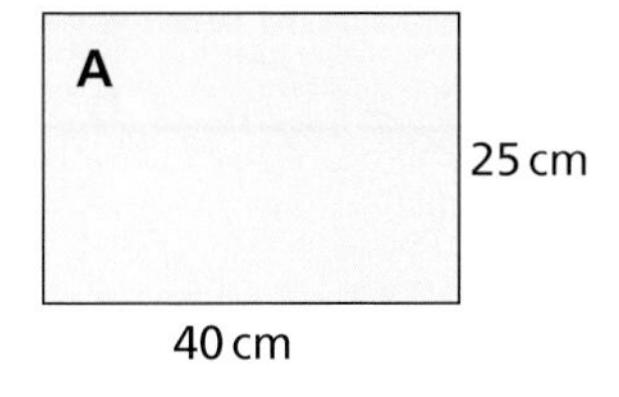

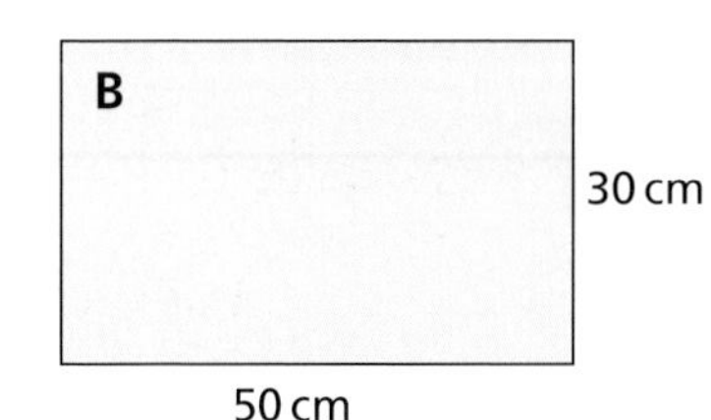

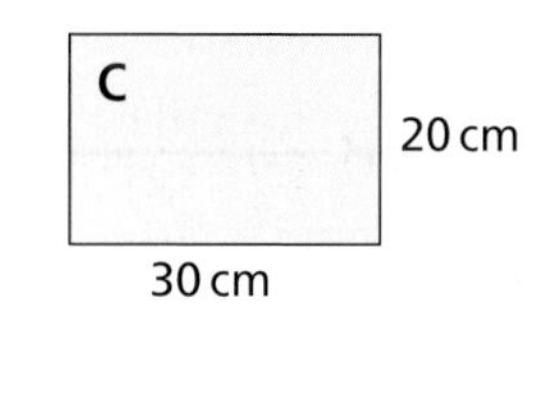

2 These spinners are both fair.
They are both spun and the two numbers are added together.

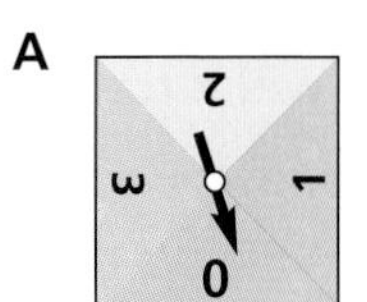

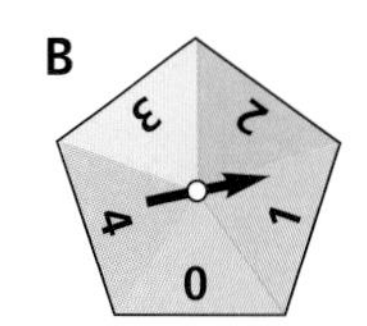

(a) Copy and complete the table of outcomes on the right.

(b) What is the probability that the total of the numbers on the spinners is less than 3?

(c) Make a similar table for the product of the two numbers.

(d) What is the probability that the product is less than 3?

(e) Find the probability that the product of the two numbers is less than their total.

		Spinner B				
		0	**1**	**2**	**3**	**4**
Spinner A	**0**	0	1	2	3	4
	1	1	2			
	2					
	3					

3 Simplify these expressions as far as possible.

(a) $3x - 4 + (x - 7)$ (b) $6x + 5 - (3x + 2)$ (c) $10x - 4 - (2x - 3)$

(d) $7x - (4 + 2x) - 9$ (e) $7 - (2 - 4x) + 5x$ (f) $8x - (5 - 2x) + x$

4 Alex did an experiment about language learning. He gave people two tests.
In test 1, people were given 20 words in an unfamiliar language and were asked to learn their meanings.
In test 2, they were given 20 Chinese characters and asked to learn their meanings.
In both cases they had three minutes in which to learn and were then tested.

Here are Alex's results.

Person	A	B	C	D	E	F	G	H	I	J	K	L	M	N	O	P	Q	R
Score in test 1	9	5	12	10	4	7	13	15	9	8	11	10	12	10	6	8	7	11
Score in test 2	5	2	8	9	4	5	8	9	4	6	8	6	7	7	3	7	6	6

Draw a scatter diagram and describe the correlation between the two test scores.

5 (a) This diagram shows part of a regular 15-sided polygon (A) and part of a regular 20-sided polygon (B).

Calculate the angles marked x and y.

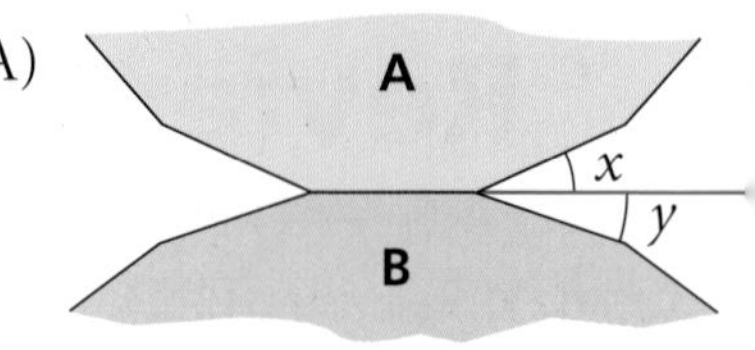

(b) This diagram shows part of a regular 12-sided polygon (C) and part of a regular 24-sided polygon (D).

Calculate the angle marked z.

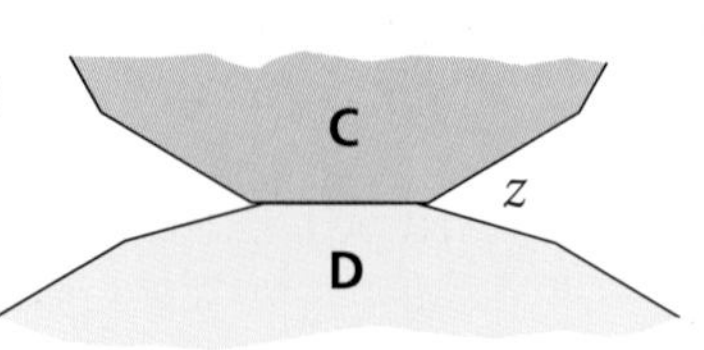

6 (a) A plane flies a distance of 325 miles in $2\frac{1}{2}$ hours. Calculate its average speed in m.p.h.

(b) The aircraft continues to fly at the same average speed. How far will it travel in 45 minutes?

(c) At 4:05 p.m. the aircraft is 210 miles from its destination. If it continues to fly at the same average speed, at what time will it arrive there? Give your answer to the nearest minute.

7 The formula $v = \dfrac{uf}{u - f}$ is used in work with lenses.

Calculate v when $u = 4.50$ and $f = 0.15$. Give the result to three decimal places.

8 Albert and Bernard live at opposite corners of a rectangular field 800 m by 500 m.

They agree to split the field so that all the points that are closer to Albert's corner than to Bernard's will be in Albert's part, and vice versa.

Draw the field to scale and draw the boundary between the two parts.

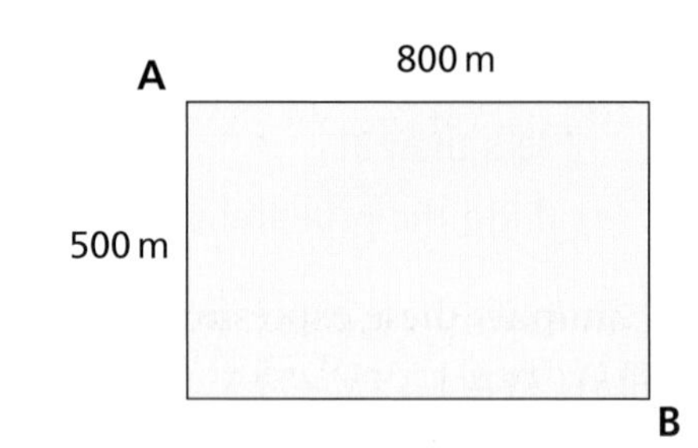

9 Forty students took an English test and a maths test. Each test was marked out of 100. This 'back to back' stem-and-leaf table shows the results. (For example, the figure picked out in red represents a mark of 27 in English.)

(a) What was the lowest mark in maths?

(b) What was the median maths mark?

(c) What was the median English mark?

(d) How can you tell from the shape of the table that the English marks are on the whole higher than the maths marks?

English		Maths
8 7 4 4 0	2	3 4 4 5 5 7 9 9
7 7 6 4 4 3 2	3	0 1 3 4 6 8 8 9 9
9 7 8 6 6 4 1 0	4	1 1 3 5 5 6 7 8
8 7 6 6 6 5 3 0 0	5	2 3 5 8 8 8
8 8 5 5 4 2	6	0 0 3 3 6
3 1 0 2	7	3 3 8
1	8	4

Stems: tens leaves: units

10 Peter, Paul and Mary keep rabbits.
Peter has twice as many rabbits as Paul, and Paul has 31 more than Mary.
Altogether they have 161 rabbits.

By forming and solving an equation, find how many rabbits each person has.

11 Each interior angle of a regular polygon is 172°.
Calculate the number of sides.

12 Arnie weighed a potato and left it in a warm place.
The potato shrivelled up and lost weight.

It weighed 127 g to start with and 78 g afterwards.
What was the percentage reduction in the weight of the potato, to the nearest 1%?

13 Mark two points A, B that are 6 cm apart.

Draw a circle of radius 8 cm with centre A and a circle of radius 4 cm with centre B.

Let P and Q be the two points where the circles intersect.
Each of these points lies on the locus of all the points that are twice as far from A as they are from B.

Draw some more circles centred on A and B with the radius from A being twice the radius from B.
Find more points on the locus.

Draw what you think the whole locus looks like.

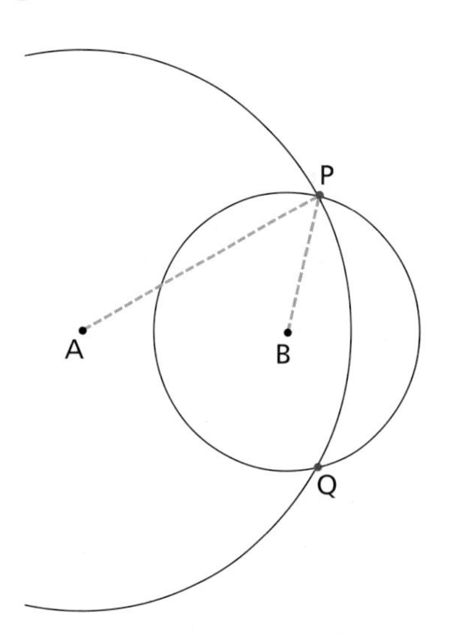

14 An ordinary cubical dice with faces numbered 1 to 6 is thrown at the same time as a regular octahedral dice with faces numbered 1 to 8.

Find the probability that the total of the two numbers thrown is more than 6.

15 The age distribution of the inhabitants of a seaside village is recorded in this table.

(a) How many people live in the town?

(b) In which age group does the median age lie?

(c) What is the mid-interval value for the 0–9 age group?

(d) Use mid-interval values to estimate the mean recorded age of the inhabitants.

Age group	Number of people
0–9	78
10–19	65
20–29	83
30–39	92
40–49	118
50–59	157
60–69	111
70–79	67
80–89	36
90–99	12

16 This graph shows the amount of fuel in the tank of a van as it is driven along a motorway.

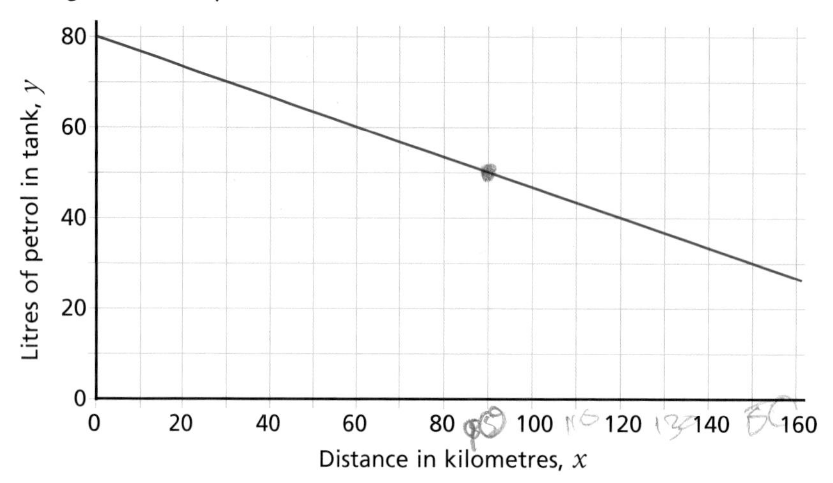

(a) Work out the gradient of the graph.

(b) Write down the equation of the graph.

(c) Use the equation to find the value of y when $x = 210$.

***17** Patti has some 5p coins and some 2p coins.
She has 129 coins altogether.
The total value of the 5p coins is 85p more than the total value of the 2p coins.

Let n be the number of 5p coins.

(a) Write down an expression for the number of 2p coins.

(b) Write down an expression for the total value of the 5p coins.

(c) Write down an expression for the total value of the 2p coins.

(d) Write down an equation that says that the total value of the 5p coins is 85p more than the total value of the 2p coins.

(e) Solve the equation and write down how many coins of each type Patti has.

***18** The houses in a street are numbered from 1 to 1000.

(a) If every house number is made from cut-out digits (for example 276), how many 5s will be needed altogether?

(b) How many 5s will be needed when the house numbers go from 1 to 10 000?

(c) Can you find a formula for the number of 5s needed for house numbers from 1 to 10^n?